FIELDBOOK

OF ILLINOIS MAMMALS

Eastern cottontail, a mammal that is common in Illinois.

FIELDBOOK

OF ILLINOIS MAMMALS

by
Donald F. Hoffmeister
and Carl O. Mohr

Dover Publications, Inc.
New York

Published in Canada by General Publishing Company, Ltd., 30 Lesmill Road, Don Mills, Toronto, Ontario.
Published in the United Kingdom by Constable and Company, Ltd., 10 Orange Street, London WC 2.

This Dover edition, first published in 1972, is an unabridged republication of the work originally published in Urbana, Illinois, 1957, as Manual 4 of the Natural History Survey Division of the State of Illinois Department of Registration and Education.
In the present edition a few errata have been rectified and several illustrations have been replaced by new ones. The frontispiece, which was originally in color, appears in black and white in this edition. A new Publisher's Note has been added.

International Standard Book Number: 0-486-20220-8
Library of Congress Catalog Card Number: 72-81283

Manufactured in the United States of America
Dover Publications, Inc.
180 Varick Street
New York, N. Y. 10014

PUBLISHER'S NOTE
TO THE DOVER EDITION

This Dover edition is a republication of the work originally published by the Illinois Natural History Survey in 1957.

In this edition minor textual corrections have been made, five figures have been replaced (figs. 42 bottom, 63, 80, 81, and 82), a running profile of the red squirrel has been substituted (facing page 1), and fig. 88 has been revised to reflect recent information on the western harvest mouse.

The western harvest mouse had been taken only in Carroll County in Illinois when this book was first published. Since that time, it has been taken in twelve additional counties in Illinois (Jo Daviess, Stephenson, Whiteside, Hancock, Stark, Peoria, Fulton, Morgan, Mason, Tazewell, McLean, and Champaign) and in Indiana just across the border from Kankakee and Iroquois Counties. The presently known distribution in Illinois is shown in revised fig. 88.

NOMENCLATURAL CHANGES

The names that have been changed since this book was originally published are listed in the left-hand column in their order of appearance in the book. The names on the right are those used today for the same animals.

Corynorhinus rafinesquii (Lesson) (pp. 39, 66, 68, 83–85)	*Plecotus rafinesquii* Lesson
Mustela rixosa (Bangs) (pp. 39, 88, 92, 97–98)	*Mustela nivalis* Linnaeus
Peromyscus nuttalli (Harlan) (pp. 40, 127, 130, 164–165)	*Ochrotomys nuttalli* (Harlan)
Myotis subulatus (Say) (pp. 67, 74)	*Myotis leibii* (Audubon and Bachman)

FOREWORD

IN 1936 the first number of the Manual series of the Natural History Survey Division appeared. It was titled the *Fieldbook of Illinois Wild Flowers*. This was followed in 1939 by the *Fieldbook of Illinois Land Snails*, and in 1942 by the *Fieldbook of Native Illinois Shrubs*. These were developed under the charge in the Illinois statute which directs the Board of Natural Resources and Conservation, under which this Survey operates, "To publish, from time to time, reports covering the entire field of zoology and botany of the State."

That these manuals have served a useful purpose is obvious in the constant demand for them. The *Fieldbook of Illinois Wild Flowers* has long been out of print, and a revised edition is now in preparation.

Now appears the fourth in this popular and useful series, the *Fieldbook of Illinois Mammals*. This work was inaugurated by Dr. Carl O. Mohr of the Natural History Survey staff over a decade ago. Upon his leaving the service of the state it was possible to persuade Dr. Donald F. Hoffmeister, Curator of the Natural History Museum of the University of Illinois and an eminent mammalogist, to continue this project. Dr. Mohr has recently returned to Illinois and has collaborated in the final stages of the project.

It is difficult for one or even two people to develop to completion a project such as this. We are greatly indebted to Mrs. Leonora K. Gloyd and Dr. Philip W. Smith of the Natural History Survey staff for the great amount of painstaking work which they expended on various duties related to the development of the manuscript. It would have been most difficult to complete the project without their unselfish assistance.

Further appreciation should be given to the Survey's Technical Editor, James S. Ayars, who has carried on his responsibilities with characteristic vigor and sensibility, to Dr. H. H. Ross, Head of the Section of Faunistic Surveys and Insect Identification of this Survey, who directed the many activities prior to the editing of the manuscript and who was responsible for continued activity over the past decade, and to Mrs. Blanche P. Young, Assistant Technical Editor, for assistance with many editorial problems.

It is hoped that this manual will be found as helpful in educational circles and among interested laymen as the three volumes which preceded it.

HARLOW B. MILLS, *Chief*

Urbana, Illinois
February 8, 1957

ACKNOWLEDGMENTS FOR ILLUSTRATIONS

Credit for illustrations used in this fieldbook should go to the following individuals and organizations:

J. C. Allen and Son, Lafayette, Indiana, for figs. 42 top, 57

American Museum of Natural History, New York, N. Y., for figs. 113, 114, 115, paintings by Charles R. Knight

James S. Ayars, Illinois State Natural History Survey, Urbana, for figs. 9, 85

William E. Clark, Illinois State Natural History Survey, Urbana, for figs. 4, 58

R. P. Grossenheider, St. Louis, Missouri, for the frontispiece, which was reproduced through courtesy of the present owner, J. Marshall Magner, also of St. Louis

E. Raymond Hall, University of Kansas, Lawrence, for figs. 61a, b, 103

N. L. Huff, Museum of Natural History, University of Minnesota, Minneapolis, for fig. 78

Illinois State Department of Conservation, Springfield, for fig. 72

Illinois State Geological Survey, Urbana, for fig. 112

Illinois State Museum, Springfield, for fig. 45

Illinois State Natural History Survey, Urbana, for figs. 11 top, 42 bottom, 43, 44, 50, 59, 63, 77, 81, 82, 86, 88, 93, 105

J. W. Jackson, Brush, Colorado, for fig. 76

Maslowski and Goodpaster, Cincinnati, Ohio, for figs. 47, 55, 64, 65, 75 bottom, 92, 94

Karl H. Maslowski, Cincinnati, Ohio, for figs. 79, 83, 90, 91, 96

Charles A. McLaughlin, University of Illinois, Urbana, for figs. 39, 51, 60, 61c, d, 73, 74

Lorus J. and Margery Milne, Durham, New Hampshire, for fig. 68

Missouri Conservation Commission, Jefferson City, for fig. 67

Carl O. Mohr, Illinois State Natural History Survey, Urbana, for figs. 1, 2, 3, 5, 6 top and bottom, 7, 8, 10, 11 bottom, 12, 14, 15, 16–20, 21–24, 25–29, 30–34, 35–36, 38, 40, 41, 61e, f, 73, 95, 107, 108, 109, 117, 118, 119, for the small drawings of chipmunk, thirteen-lined ground squirrel, Franklin's ground squirrel, red squirrel, and eastern gray squirrel in the front of this book and on pages 212 and 222, and for the drawings on pages 218 and 219

Ernest E. Mulch, Phoenix, Arizona, for fig. 54

Museum of Comparative Zoology, University of California, Berkeley, for fig. 80

Natural Resources Council, Conservation News Service, Washington, D. C., for fig. 69

New York Zoological Society, New York, for figs. 66, 87
Gordon Pearsall, formerly Forest Preserve District of Cook County, River Forest, Illinois, for fig. 62
Herbert H. Ross, Illinois State Natural History Survey, Urbana, for fig. 98
Walter J. Schoonmaker, New York State Museum, Albany, for fig. 75 top
Charles and Elizabeth Schwartz, Jefferson City, Missouri, for fig. 70
Charles L. Scott, formerly Illinois State Natural History Survey, Urbana, for fig. 89
Thomas G. Scott, Illinois State Natural History Survey, Urbana, for fig. 13
United States Fish and Wildlife Service, Washington, D. C., for figs. 37, 71, 97, 104
United States Public Health Service, Communicable Disease Center, Atlanta, Georgia, for figs. 99, 100 bottom
University of Illinois, Department of Forestry, Urbana, for fig. 116
University of Missouri, Extension Service, Columbia, for fig. 106
Ernest P. Walker, Smithsonian Institution, Washington, D. C., for figs. 46, 48, 49, 52, 53, 56, 84, 101, 102
Wisconsin Alumni Research Foundation, Madison, for fig. 100 top
Wisconsin State Department of Conservation, Madison, for figs. 110, 111

CONTENTS

INTRODUCTION

MAMMALS characteristically are warm-blooded animals that have backbones, are at least partly covered with hair of some kind during their lives, possess mammary (milk) glands, and give birth to living young. Any animal occurring wild in Illinois and having four limbs and a coat of fur or hair is easily recognized as a mammal.

In addition to man and his domesticated animals, 59 species of mammals are known to occur in Illinois. Of these species, three—roof rat, Norway rat, and house mouse—were introduced from the Old World. The others are native. The commonest Illinois mammals are mice, shrews, rabbits, and squirrels. Larger mammals, such as beavers, badgers, and deer, are less abundant. Bison, elk, bears, wolves, mountain lions, and some other mammals that once roamed the prairies or lived in the forests of the Illinois area have vanished from the state, or are present only as caged animals.

In pioneer days, many wild mammals served as sources of food and clothing for settlers and as means of livelihood for trappers. Although they are no longer of paramount importance in this respect, the wild mammals of Illinois today continue to be a significant asset. The furbearers, such as muskrats, minks, and foxes, have yielded annually more than a million dollars worth of fur. The game animals, such as rabbits and squirrels, have provided millions of pounds of food each year. Also, some of the wild mammals are a source of welcome recreation to hunters and naturalists.

WHERE MAMMALS LIVE

The casual observer or the average hiker sees very few wild mammals out-of-doors and he may get the impression that they are a rarity. Actually, mammals are numerous in almost every acre of uncultivated land in Illinois—prairie, forest, meadow, fencerow, and marsh—probably 5 to 12 times as numerous as birds. In woods there may be 10 to 60 mammals per acre; in a meadow, along fencerows, and in grassy places along forest

1

Silver-haired bat	Flying squirrel		Raccoon
	Fox squirrel	Deer	
		Gray fox	
Opossum	Mole mound		Chipmunk
White-footed mouse	Masked shrew		Short-tailed shrew
Mole	Pine vole		

Fig. 1.—Characteristic mammals of the unpastured Illinois forest.

edges, the number per acre may be even greater. Of this number, about 95 per cent are small mammals—chipmunk size or smaller. As the hiker wanders through a forest or crosses a brushy field, he may walk near hundreds of these small animals without being aware of them.

Habitats.—Each type of habitat has its characteristic denizens. Under the forest floor may be shrews, moles, and voles. On the forest floor may be skunks, raccoons, foxes, opossums, white-footed mice, and chipmunks. Above the forest floor may be flying squirrels, tree squirrels, and various kinds of bats. By probing into the leaf mold on the forest floor, one may find the intricate maze of tunnels used by shrews and voles; by breaking open or turning over a decaying log, one may discover the nest of a white-footed mouse or a pine vole; or, by probing into an old woodpecker hole high in a tree, one may arouse a flying squirrel from his sleep. Mammals characteristic of the Illinois forest are shown in fig. 1.

In a meadow or fencerow may live one or more coyotes, red foxes, striped skunks, cottontails, woodchucks, weasels, and smaller animals. If one looks closely in the grass, he may find signs of the prairie vole, the least shrew, the thirteen-lined ground squirrel, and the deer mouse. Some mammals characteristic of this habitat are shown in fig. 2.

In houses built in the water of lakes, marshes, or streams live muskrats and beavers. At the edges of streams and lakes, one may see tracks of raccoons, minks, and muskrats. Some mammals characteristic of this habitat are shown in fig. 3.

Many wild mammals are active principally at night; hence, most people are unaware of their presence and unfamiliar with the places in which they live. The comparatively large nocturnal mammals—as raccoons and opossums—are occasionally encountered by day, but the more abundant, small nocturnal forms such as shrews and mice are seldom seen. Most squirrels are active and above ground during the day and for this reason are seen more often than most other wild mammals. Fox squirrels in the woods are obvious during the daytime as they hunt among leaves on the ground; but flying squirrels, which may be just as numerous, are seldom seen because normally they sleep all day. Shrews, voles, and other mice are easily overlooked because they spend much of their lives in their burrows and nest chambers.

Skunk   Red fox   Little brown bat
        Cottontail   Woodchuck
Pocket gopher mound
                Thirteen-lined ground squirrel
        Least shrew   Prairie vole   Deer mouse
        Pocket gopher   Storage chamber of prairie vole

Fig. 2.—Characteristic mammals of Illinois fields and fencerows.

Golden mouse and nest   Raccoon
Mink
Cotton mouse Swamp rabbit droppings on log
Rice rat

Bobcat
Muskrat and lodge
Swamp rabbit

Fig. 3.—Characteristic mammals of Illinois swamps and stream margins.

Geographical Distribution.—Each species of mammal is able to exist only in certain types of habitat and under certain climatic conditions. The area in which a species exists is called its range or distribution pattern. The range for each species may change as the habitat or other conditions change.

Enough collecting has been done to provide a knowledge of the approximate range of each species of Illinois mammal, but for many species the details of distribution are not well understood. A knowledge of the range of the species is valuable in giving clues concerning climatic or habitat conditions which control its distribution.

The study of the geographic distribution of mammals is extremely interesting, and amateur collectors have a real opportunity to make worth-while contributions to this study. The value of their contributions, however, depends on the care and accuracy with which they collect, preserve, and label their material. Instructions for the collector are given in this book, and a general knowledge of the ranges of North American mammals can be gleaned from such standard mammalogy books as *A Field Guide to the Mammals* by Burt (1952) or *The Mammal Guide* by Palmer (1954). The collector can check his records against the maps and accounts of distribution in these books as well as in the following pages to see if he has records of interest.

Species State-Wide in Occurrence.—Of the 59 species of wild mammals in Illinois, 36 occur throughout the state or in localities scattered throughout the state. These 36 species include some whose ranges center in Illinois and others whose ranges center to the north, south, east, or west of this state. Species such as the cottontail and the house mouse are abundant and of state-wide distribution in Illinois; others such as the white-tailed deer and the beaver are also state-wide in distribution but are comparatively rare and sporadic in their occurrence.

The distribution pattern and abundance of the various species of mammals reflect the occurrence, extent, and quality of habitats suitable for their existence. For example, chipmunks occur in many localities in Illinois but are confined to wooded, ungrazed hills and ravines. Moles are generally distributed in Illinois but, as they require well-drained soils, few of them are found in the marsh and bog areas of the extreme northern part of the state.

Species Not State-Wide in Occurrence.—Four kinds of mammals—the masked shrew, pigmy shrew, least weasel, and meadow vole—are essentially northern animals and seem to be limited in Illinois to localities within the northern half of the state. The red squirrel, which may still occur in Illinois, although it is not listed among the species known to be in this state, is also a typically northern mammal and, if present, would be expected only near the northern border.

Ten species of Illinois mammals are primarily southern in distribution. Six of them—the southeastern bat, big-eared bat, golden mouse, cotton mouse, rice rat, and wood rat—occur in the southern fourth of the state. Two of them—the gray bat and the swamp rabbit—occur a little farther northward, and two others—the evening bat and the eastern pipistrel—occur over most of Illinois.

Six Illinois species are predominantly western in distribution. If only the Illinois distribution of these mammals were considered, they might be regarded as northern animals. However, these animals are prairie or plains species and follow the western prairie habitats which extend eastward across the northern half of Illinois. Three species—the badger, the thirteen-lined ground squirrel, and the Franklin's ground squirrel—occur in Illinois in only the northern half, but two species—the western harvest mouse and the white-tailed jackrabbit—are restricted in this state to the northwestern corner. The sixth western species—the plains pocket gopher—is unique among Illinois mammals in that the remnant colony in the state is now geographically isolated and is distinct in color from its closest relatives. This colony occurs in a few counties of Illinois and adjacent counties of Indiana, in an area bounded on the north by the Illinois and Kankakee rivers.

For several species of mammals our knowledge of their total range or their range in Illinois is scant. One of these species is the introduced roof rat, which is common in the South but which in the North is found sporadically in only a few areas, usually in cities. Another species, the Indiana bat, is known from a very few scattered localities, and until more records become available we will have no clear picture of its range. The southeastern shrew also has a poorly known distribution; it is recorded from southeastern United States and from a few localities in Illinois and Indiana.

HOW TO STUDY MAMMALS

Mammals may be studied alive or as preserved specimens.

Studying Live Mammals.—Live mammals may be studied (1) directly, in the field or in captivity, and (2) indirectly, by means of their signs—homes, trails or runways, tracks, droppings, and food fragments or remains. Both types of observations are of importance in contributing to an understanding of the habits of the animals.

To study mammals in their normal activities in the field, one must approach them slowly and inconspicuously, remain hidden in a spot to which they come, or attract them by artifice. One can sometimes attract mammals by placing baits at strategic spots or by making imitative calls. For example, an appropriate squeaking sound will sometimes bring an inquisitive chipmunk within range for easy observation. At night one can follow the activity of some mammals in the beam of a flashlight or in the light of a gasoline lantern.

One good method of studying small mammals and their life histories is as follows: Select a plot of grass or woodland in which mammals are present and mark it into quadrats; make a record on cross-section paper of the runways, nests, and feeding stations as they occur in each quadrat; keep notes on the number of individuals living on the plot, time of activity, presence and development of young in the nests, and any other items you may notice. Ground squirrels, chipmunks, and voles are particularly amenable to such observation.

Another method of studying mammals is to observe them in captivity. Care should be taken to provide quarters that are suitable for the animals and for ease of observation. For some of the shy and nocturnal mammals this type of study may be one of the most efficient ways of obtaining accurate information on breeding, rearing of young, food preferences, and behavior characteristics. The animals can be caught in live traps, of which there are many kinds to be purchased or made, set near runways or nesting sites. Mice or shrews may be trapped in a gallon can or other steep-sided container that has been sunk in the ground flush with a runway. Some mammals that live under logs or other objects on the ground can be captured by raising the objects and seizing the animals before they dash away.

The number of wild mammals that can be observed in an area may be increased by improving the habitat for them. Where there is a scarcity of suitable places for nesting and hiding, and little available food, the mammalian population is very low. In agricultural areas, cultivation may be so intensive that few places are left that are suitable for mammals. Sometimes a few changes in these situations will increase the number of mammals which can live there. In a wood lot having few hollow trees, the addition of some nest boxes fastened in the trees may provide homes for raccoons, squirrels, and opossums. A fence of multiflora roses should provide sanctuary for cottontails and mice. A ditchbank on which grass and other vegetation are allowed to flourish will produce many more muskrats than one that is grazed, mowed, or denuded. The removal of brush and rotting logs from a wood lot destroys the habitat of mice and squirrels and, indirectly, of other animals that feed on these small mammals.

Making a Collection of Preserved Specimens.—If one wishes to make a survey of the mammals occurring in an area, he should collect and prepare specimens of at least the small animals so that they may be identified accurately. Securing specimens of the various species is usually accomplished by hunting or trapping. Snap-traps are best for trapping small animals. Some specimens can be found in steep-sided excavations such as dry cisterns, window wells, post holes, and deep ditches into which they have fallen. Often, revisits to such places are profitable because animals of different species may fall in at different times. Mammals as large as or larger than rats may be shot or trapped. Because most of them are protected by law, special permission must be obtained unless they are collected only during the hunting season, when they may be taken under an ordinary hunting or trapping license. Highways may yield some specimens that have been killed, but not badly mutilated, by cars.

A specimen that is small may be prepared for a study collection in one of the following ways. If only the skeleton is desired, the animal is skinned, eviscerated, and dried; later it may be cleaned by a culture of dermestid beetles. If for special reasons preservation of the whole animal is desired, it can be kept in 10 to 15 per cent formalin in a glass jar. Newly born young are often preserved in this manner. The recommended method

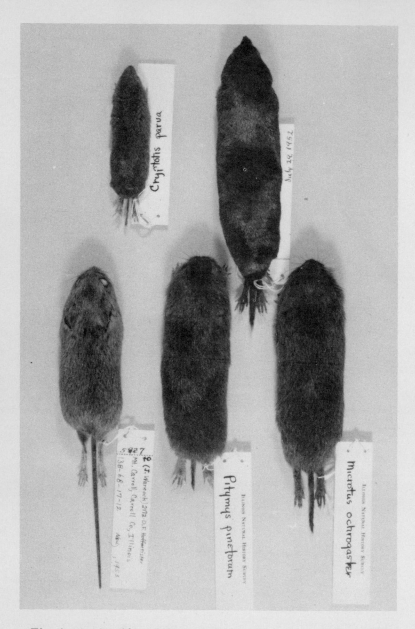

Fig. 4.—Study skins: left to right, top row, least shrew, short-tailed shrew; bottom row, western harvest mouse, pine vole, prairie vole.

of keeping most specimens for scientific study is as dry skins, fig. 4, and skulls. After a little practice, such skins can be prepared by a simple type of taxidermy. The following suggestions give some idea of the steps involved in making a collection of study skins of small mammals.

1. Assign each specimen a collection number; use this same number for the skin, the skull, and any field notes or other data that may have been recorded for the animal.

2. Record in a catalog or notebook the full collecting data (precise locality, date, habitat, collector) and any additional notes you may wish to make.

3. Determine the sex and take the measurements of the animal before starting to prepare it for a study skin or skeleton and record them in the catalog or notebook; also write them on a tag (see step 6 below). Usually the measurements taken are over-all or total length, length of tail, length of a hind foot, and length of an ear. The head-and-body length is usually not included in the standard measurements that are written on the tag in the following sequence: total, tail, hind foot, ear. The weight of the animal also is sometimes recorded on the tag.

The over-all or total length is the distance from the tip of the nose to the tip of the tail, not including the hair at the end of the tail. Tail length is the length of this appendage, exclusive of hair at the tip, when it is at a right angle to the body axis. Hind foot length is the distance from the back of the heel to the tip of the longest claw of a hind foot; only one foot is measured. Ear length is the distance from the notch of an ear to the tip of the pinna or projecting part of the ear; the length of any tuft or other hair is not included. Usually and preferably, measurements are given in the metric system and in millimeters. One inch is roughly equal to 25 (25.4) millimeters (written mm.) and 1 foot is equal to 305 millimeters.

4. Prepare the study skin. Make an incision down the venter of the mammal from the mid-thoracic region to a point between the hind legs; then skin out the body, legs (to the knee or elbow joints or below), and head. Remove the tail bone from the sheath of skin covering it, much as you would slip an ice pick from its leather holder. Carefully clean fat and grease from the inside of the skin so as to prevent future staining and corrosion of the hair side. Fill the skin with a carefully fashioned, firm cylinder of cotton batting designed to approximate original con-

tours of the animal. Before inserting the cylinder of cotton batting, shape the front end of it into a cone somewhat like the configuration of the head. Place rustless, annealed or tinned wire, partly covered with cotton, in the skin to replace or supplement bones of the tail and the legs and feet. Then sew up the skin along the original incision, pin the specimen in the desired position to a sheet of cork or corrugated cardboard, and allow it to dry.

5. Save and clean the skull. Attach to it a tag bearing the same number as that given to the skin, so that the two can always be associated.

6. To the completed skin, attach a label giving the collector's name and the collection number, the precise collecting place, date, sex, and standard measurements of the specimen; allow a space for filling in the name of the species when the specimen has been identified.

A specimen that is large may be prepared either in approximately the same manner as a small one or as a cased or flat skin. An incision is made along each back leg from the base of the toes to the anus; then the pelt is peeled forward over the body—that is, it is cased. All bits of fat and flesh should be removed. The pelt is then fitted over a board or frame that is rounded at the front end; it is then allowed to dry. A thoroughly cleaned and dried pelt will keep for months if stored in a cool place, but tanning is required for long-time keeping qualities and to reveal the beauty of a good pelt. A flat skin is prepared in the same way as a cased skin except that additional incisions are made from the anus to the mouth and from the base of the toes on one front leg, across the chest to the base of the toes on the other front leg so that the skin can be pinned out flat to dry. A pelt that is to be tanned should be pinned with the fur side next to the drying board to allow for rapid drying.

There are many details and precautions to be taken in measuring and putting up mammal skins. The University of Illinois Museum of Natural History has available a motion picture showing how to prepare specimens. The following books describe the skinning and stuffing techniques: *Methods of Collecting and Preserving Vertebrate Animals,* by R. M. Anderson, 1948, Bulletin 69, National Museum of Canada; *Handbook of Mammals of Kansas,* by E. R. Hall, 1955, Miscellaneous Publication 7, University of Kansas Museum of Natural History; and *A Field Collector's Manual in Natural History,* 1944, Publication 3766, Smithsonian Institution, Washington, D. C. The beginning collector should study these books carefully.

SIGNS

The study of mammal signs is one of the most interesting pursuits available to a hiker. By practice in developing an eye for details, the most observant hunters, trappers, and naturalists are usually highly skilled in reading signs. The occasional hiker can train himself to discern and interpret signs and thus add considerable interest to his field trips.

Most mammal signs fall into one of these six classes: homes, trails and runways, tracks, scats or droppings, tooth marks, and food stores and fragments. The interpretation of these signs is a type of detective work, and it is largely a process of elimination combined with a general knowledge of the mammals occurring in the region. Several factors, such as the kind of sign, size and form of the sign, ranges of mammals which could have made such a sign, type of habitat, and season of the year, must be considered in each case. For example, suppose the sign consists of a set of small tracks in a snow-covered field adjacent to a tree-lined ditch in southern Illinois. The size of the tracks hint that the mammal is smaller than a cat but larger than a rat, thus eliminating many species. The fact that ground squirrels, pocket gophers, and red squirrels do not occur in southern Illinois removes them from consideration. The habitat suggests that the animal is a forest-edge species, a prairie animal, or even an aquatic species, but probably not an inhabitant of deep woods, such as is the gray squirrel. The snow on the ground precludes the possibility of a woodchuck, because this animal would be in hibernation. The list of possible mammals still includes the fox squirrel, muskrat, mink, weasel, and cottontail. A close scrutiny of the tracks reveals that the toe marks are long and slender, thus ruling out the cottontail, weasel, and mink, but leaving the fox squirrel and the muskrat. The tracks can be compared with sketches of muskrat and fox squirrel tracks, figs. 21 and 34; these sketches indicate that the muskrat is likely to leave some trace of a tail mark. If a tail mark is present, it is likely that the tracks were made by a muskrat.

Fig. 5.—Entrance to burrow of woodchuck.

Fig. 6.—Signs of eastern mole: top, runway; bottom, mound.

This hypothetical case illustrates the type of reasoning that may be employed in identifying signs. In many cases the list of suspected mammals can be reduced to a few kinds but not to one specific mammal. If a combination of signs is available, such as homes and tracks or runways and food fragments, there is a good chance for specific diagnosis. In some cases, signs can be attributed to a specific animal with certainty. Indirect clues, such as range, habitat, and season, are valid only if based on an excellent knowledge of the animals likely to be present.

Homes.—The homes of Illinois mammals fall into two chief categories: burrows, or homes in the ground, and homes on or above the ground.

Some kinds of Illinois mammals are adaptable in habits, living either in burrows or in homes above ground; a few kinds may inhabit a variety of types of homes. Most species, however, show a decided preference for particular types of homes.

Burrows.—Mammals inhabiting burrows include those that dig their own burrows and those that use burrows made by other kinds of mammals. The size of the entrance, size and depth of the nest below ground, and appearance of the burrow are indicative of the kind of mammal living in a burrow.

Burrows can be classified, according to size, as follows:

1. Diameter less than 2 inches.—Characteristic of about a dozen species of mice and voles, the shrews, and small rice rats.

2. Diameter 2 to 4 inches.—Characteristic of ground squirrels, gophers, chipmunks, moles, and barn rats.

3. Diameter more than 4 inches.—Used by large mammals such as woodchucks, fig. 5, badgers, skunks, opossums, raccoons, otters, and coyotes.

Weasels and minks commonly take over homes of other species. A small weasel may appropriate a mouse nest and a mink may occupy a woodchuck den.

The size of the burrow may indicate the size of the mammal. The location of the burrow, the manner in which excess dirt is deposited, or other signs at the den site may furnish additional identifying marks. A large hole with a muddy slide nearby may belong to an otter, a large one with feathers and rabbit remains may be the den of a fox, and one with a musky odor may be the home of a skunk or badger.

Among the medium-sized burrows and soil dumps the following are distinctive:

1. A steep mound of earth, usually circular and conical, thrust upward from the center and associated with one or more subsurface runs which hump up the ground and are readily visible, fig. 6, is typical of the work of moles.

Fig. 7.—Mound of plains pocket gopher.

Fig. 8.—Entrance to burrow of Norway rat.

Fig. 9.—Arboreal nest of eastern fox squirrel.

2. A more irregular mound, fig. 7, lacking the visible indication of subsurface runs and the symmetry of the mole mound, is typical of pocket gophers. This is usually a rough crescent with a small pustule midway between the tips of the crescent.

3. A hole with a mound of earth to one side and trash scattered about, fig. 8, is typical of Norway rats.

4. A hole in a grassy area with a flat bare area to one side is typical of thirteen-lined ground squirrels.

There are, of course, many variations from the typical burrows described above.

Homes Above Ground.—Many species of mammals make grassy or leafy nests on the ground or in shrubs or trees. Mammal nests are almost invariably roofed over, not open as many birds' nests are, and they do not have the nest materials plastered together with mud or secretions. However, certain mammals reconstruct birds' nests for their own homes. For this reason the only sure way to ascertain the inhabitant of a covered nest is to see the inhabitant as it enters or leaves the nest or to examine the nest for the presence of eggs, young, or an adult.

Mammal homes above ground may be divided conveniently into three categories: arboreal, hidden terrestrial, and aquatic.

Arboreal nests include those made in trees, in bushes, or in tufts of herbaceous plants. Several distinctive types may be found, as follows:

1. Nests made in hollows of trees. These may be utilized as home sites by raccoons, opossums, fox squirrels, or gray squirrels, flying squirrels, bats, white-footed mice, and even gray foxes or bobcats.

2. Large exposed nests high in trees. These may be the homes of tree squirrels, fig. 9; if deserted by the original owners, they may be occupied by raccoons, bats, or sometimes flying squirrels. At a distance these nests look like crows' nests but they may be distinguished by being completely roofed over and usually being composed of leaves rather than of sticks and leaves.

3. Smaller nests in bushes or herbaceous plants near the ground. These nests are made chiefly of dry grass or of plant "down." Many small mammals have nests of this type. These nests may be entirely the work of the mammals occupying them or they may be reconstructed birds' nests. Such homes are characteristic of golden mice, fig. 91, cotton mice, white-footed mice, and rice rats.

Hidden terrestrial homes are those made in brush piles or rock crevices, or beneath logs or other objects. Many species of mammals make homes of this type.

Skunks occasionally make homes under the flooring of old buildings. Several species of rats, mice, and shrews construct nests under logs, and opossums and gray foxes sometimes make their homes inside hollow logs. Rabbits and long-tailed weasels often make homes in brush piles. Hares hollow out pockets in

Fig. 10.—Form of white-tailed jackrabbit.

the ground, called forms, fig. 10, each just deep enough to partially conceal the animal occupying it.

Wood rats, white-footed mice, chipmunks, gray foxes, raccoons, and bobcats frequently make nests or dens in fissures and crevices of rock bluffs.

Aquatic homes are those built in and above the water of marshes, ponds, and streams. Muskrats and beavers make lodges, fig. 11, and burrows, fig. 12. If these homes are abandoned by the animals making them, they may be taken over by minks.

High Trails, Runways, and Slides.—A few Illinois mammals make trails that are conspicuous and characteristic of the animals making them. Foxes, coyotes, and woodchucks occasionally beat trails through dense vegetation, fig. 13, especially in the vicinity of their burrows. Voles make distinctive trails through grass; these trails frequently are exposed by grass fires in spring, fig. 14. Muskrats and minks in swimming through

Fig. 11.—Top, lodge of muskrat; bottom, lodge of beaver.

duckweed beds leave open-water trails which generally are longer than those left by fish feeding at the surface. Most mammals, however, use trails common to several species, or use avenues of travel not recognizable as definite trails. For example, a stream edge may exhibit many tracks of raccoons, opossums, and minks, indicating a great deal of night traffic but no well-defined land trail. A forest edge may serve as an avenue of travel by many species, although it may have no signs to identify it as a trail.

Recognizable trails may be classified as high trails, runways, or slides. There are often no sharp distinctions between these three types, and some trails are difficult to assign to any one type.

High trails are open paths, usually without canopies of grasses or bushes, and they usually belong to relatively large animals. Bison trails and deer paths are of this type and formerly were common in the Illinois area. Present-day mammals making high trails are red foxes, fig. 13, rabbits, woodchucks, and coyotes.

Fig. 12.—Burrow and runways of muskrat.

Runways are tunnels or small trails that often are so low that the grass must be parted or the debris scraped away if they are to be discerned by human beings. Although they are not often seen, runways are abundant in both wooded and grassy areas. They may be classified as those that are on the surface of the ground, fig. 14, and those that are just under the surface. Many runways that appear to begin on the surface disappear below the surface and then reappear, forming a complicated maze. Surface runways are much the same for a dozen or more kinds of shrews, voles, and other mice, but the habitat may provide clues for identifying the species that use them. For example, runways in the debris of a forest floor are usually

Fig. 13.—High trail of red fox.

those of pine mice, shrews, or white-footed mice. Similar runways in a meadow or pasture are more likely to be those of
voles, bog lemmings, or deer mice. Other signs such as scats or

Fig. 14.—Top, mound, and, bottom, surface runways of prairie
vole. Normally these are hidden in the grass but in the cases
shown here the grass has been burned away; the mound and the
maze of runways that lead to holes opening into a complex underground system of burrows are exposed.

food fragments may be available to aid in eliminating some species from consideration.

Slides are worn troughs on muddy stream banks where otters, minks, or muskrats have slid into the water.

Tracks.—Tracks are among the most reliable signs for identifying mammals, but their characteristics must be memorized or the tracks compared directly to sketches of known kinds of tracks. Tracks of many mammals native to Illinois are illustrated in the accompanying plates. With a little practice in identifying tracks, a person can learn to recognize many of these at a glance.

Kinds of Tracks.—The kind of track a mammal makes depends upon whether it is walking, trotting, or running. Tracks made by an animal when running have individual footprints farther apart than those made by the same animal when walking, and often the footprints are arranged in a different pattern.

In mud, moist sand, or snow, a mammal may make excellent imprints of its feet and sometimes belly or tail. When made in thick mud, the prints of the toes are usually spread more widely than when made in a layer of thin mud. If distinctly made tracks are found in Illinois, they may be identified through use of the following key.

Key to Tracks of Some Common Illinois Mammals

1. Footprint consisting of one or a pair of solid impressions, figs. 15, 16 2
   Footprint consisting of three or more toe marks, figs. 17–30 6
2. Each footprint less than 1½ inches wide 3
   Each footprint more than 1½ inches wide 4
3. Each set of 4 footprints with a pair tandem and a pair side by side, or almost side by side, fig. 16 . . . . rabbits
   Each set of 4 footprints arranged otherwise . . . .
   . . . . imperfect tracks of small mammals
4. Each footprint a single subcircular mark . . . . horse
   Each footprint consisting of a pair of marks, fig. 15 5
5. Each footprint 3 inches or more across . . . . cow
   Each footprint less than 3 inches across, fig. 15 . . . .
   . . . . deer, goat, sheep, pig, calf
6. Imprint of first and of fifth toe of front foot at almost 180-degree angle, fig. 17; axis of imprint of big toe of hind foot well separated from and at an angle of at least 90 degrees to axis of imprint of next toe . . . . opossum
   Imprint of inner and of outer toe of front foot at less than 120-degree angle, figs. 18–36; imprint of inner and of outer toe of hind foot at less than 90-degree angle 7

7. Each set of 4 footprints with a pair tandem and a pair side by side or almost side by side, fig. 16 . . . . rabbits
   Each set of 4 footprints usually arranged otherwise 8
8. Claw marks absent or continuous with the marks made by the toepads, figs. 18, 21 9
   Claw marks usually apparent and separate from marks made by the toepads, figs. 24, 29 10
9. Print of hind foot with 4 toe marks, figs. 18, 25 . . . .
   . . . . weasels, mink, cats
   Print of hind foot with 5 toe marks, figs. 20–23 . . . .
   . . . . raccoon, muskrat, rats, river otter
10. Print of hind foot with 4 toe marks, figs. 24, 26, 27 . . . .
    . . . . dog, coyote, foxes
    Print of hind foot with 5 toe marks 11
11. Footprints showing webbing between toes, fig. 28 . . . . . . beaver
    Footprints not showing webbing, fig. 29 12
12. Prints of front and hind feet rounded, figs. 29–33 . . . .
    . . . . woodchuck, mice, voles, shrews
    Print of hind foot elongate, figs. 20, 22 13
13. Rounded print of front foot with 4 toe marks, figs. 22, 34 . .
    . . . . squirrels, rats, mice
    Rounded print of front foot with 5 toe marks, figs. 20, 35, 36
    . . . . raccoon, skunk, badger

Preserving Tracks.—Tracks can be preserved for study in four ways. The simplest way is to photograph imprints in snow or mud and make a collection of pictures of the different types. A second way is to sketch the imprints and keep the sketches in a notebook or card file. A third way, which is interesting but somewhat more laborious, is to make plaster facsimiles. This may be done by pouring plaster of Paris mixed with water into and around the tracks and allowing it to harden; then removing the plaster cast, brushing it clean, greasing it, and pressing it into a plate of freshly mixed plaster, thus duplicating the original impressions. A fourth way, similar to the third, can be used for tracks in mud by letting tallow from a burning candle drip into the imprints. When the tallow solidifies, it can be lifted up, cleaned, and stored, or used to make a plaster cast.

Scats or Droppings.—The study of scats or droppings (scatology) is often helpful in providing clues to the identity of the mammals frequenting an area and also in determining the food habits of these mammals. As in the case of tracks, some scats are easily recognized, others have characteristics which seem to defy adequate description but can be learned by observation, and some cannot be identified with certainty even by experts.

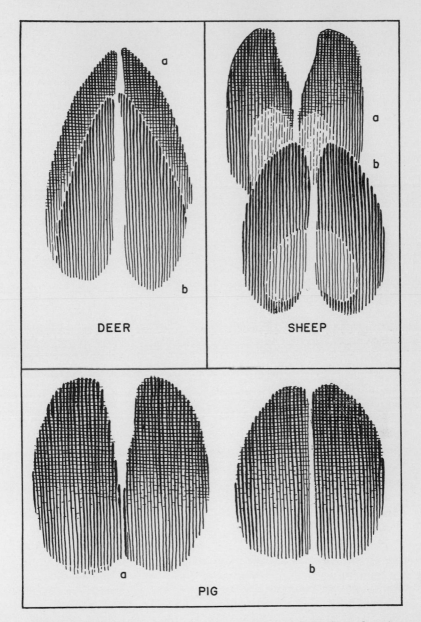

DEER   SHEEP

PIG

Fig. 15.—Hoof prints of ungulates: *a*, print of front foot; *b*, print of hind foot.

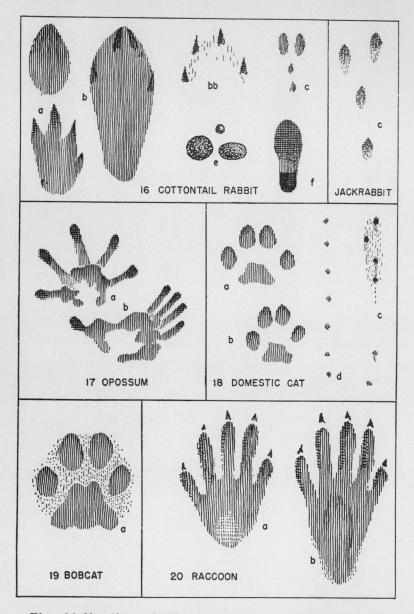

Figs. 16–20.—Signs of Illinois mammals: *a,* print of front foot; *b,* print of hind foot; *bb,* claw marks of hind foot; *c,* bounding pattern; *d,* walking pattern; *e,* droppings; *f,* shoe print, scale of 16*c.*

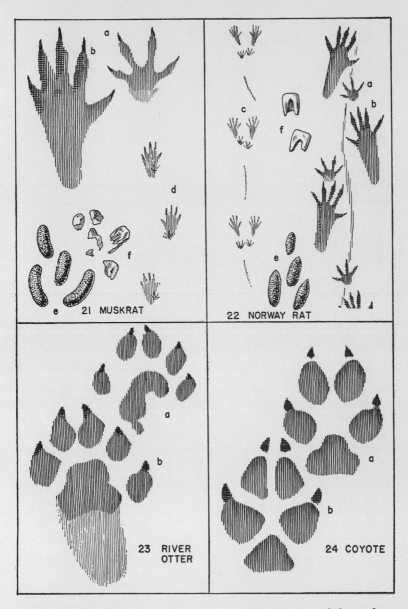

Figs. 21–24.—Signs of Illinois mammals: *a*, print of front foot; *b*, print of hind foot; *c*, bounding pattern; *d*, walking pattern; *e*, droppings; *f*, food fragments.

Figs. 25–29.—Signs of Illinois mammals: *a,* print of front foot; *b,* print of hind foot; *c,* bounding pattern; *d,* walking pattern; *e,* tooth marks on branch.

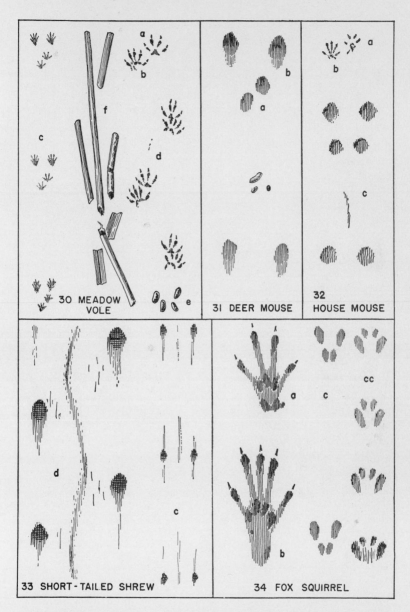

Figs. 30–34.—Signs of Illinois mammals: *a*, print of front foot; *b*, print of hind foot; *c*, bounding pattern; *cc*, hop and stop pattern; *d*, walking pattern; *e*, droppings; *f*, food fragments.

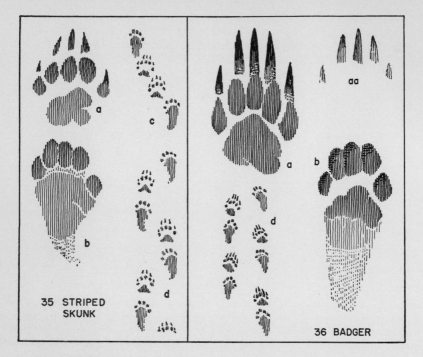

Figs. 35–36.—Signs of Illinois mammals: *a*, print of front foot; *aa*, claw marks of front foot on hard surface; *b*, print of hind foot; *c*, bounding pattern; *d*, walking pattern.

Scats of carnivores and opossums are cylindrical objects which contain mats of hair, feathers, sometimes teeth or bone fragments of vertebrates, and parts of insects. Those of omnivorous mammals often contain seeds of fruits mixed with the animal remains. Scats of most herbivores are small pellet-like or oval objects, usually present in considerable numbers. Often the place of deposition of scats is an important distinguishing characteristic; for example, scats of the swamp rabbit are in appearance practically indistinguishable from those of the cottontail, but the swamp rabbit has a habit of leaving droppings on logs or mounds, whereas the cottontail leaves droppings on level ground.

Tooth Marks.—In obtaining food or in constructing runways or homes, certain mammals leave distinguishing tooth marks or signs of gnawing. Rabbits and voles commonly, muskrats and other mammals occasionally, gnaw and eat the bark of trees, fig. 37. Rats and mice gnaw woodwork in buildings, fig. 100, food packages, and other materials in constructing run-

Fig. 37.—Winter damage to fruit tree by rabbits and voles. The damage to the branches is by rabbits, that to the lower part of the trunk by voles.

Fig. 38.—Feeding platform of muskrat. The rice rat makes a feeding platform that is smaller and of finer vegetation.

ways or nests or in feeding. Beavers, in their feeding, leave marks of their teeth on branches, fig. 28e, and, in their home building, fell and cut up trees, even large ones.

Food Signs.—Food caches, partially eaten food, certain unpalatable objects, and discarded fragments of food items may furnish clues to the mammals associated with them. Food fragments may be either animal remains or plant remains.

Animal Remains.—Wings of flies and moths on the floor of a building or cave, or within a hollow tree, are excellent signs of bats. Insect remains mixed with snail shells in a ground nest usually mean the presence of shrews. A shallow hole with fragments of turtle eggs scattered about usually means that a skunk or a raccoon has had a feast there. Bird feathers around a hole in a bank suggest the presence of a mink or weasel, and bird feathers and rabbit remains around a large hole in an upland situation indicate the presence of a fox or coyote.

Vegetable Remains.—Bundles of cut grasses on the open ground, each section 2 to 3 inches long, indicate that a jumping mouse lives nearby. If the sections are shorter, fig. 30, and in a runway, they suggest the work of voles, or their relatives the bog lemmings or pine mice. In a rock crevice, a cache of nuts and seeds with such inedible objects as paper and corncobs hints of the presence of a wood rat; in or near human habitations, these signs suggest the activity of a Norway rat or a house mouse. Platforms, fig. 38, of freshly cut grass, cattails, or rushes in shallow water indicate the feeding site of a rice rat or a muskrat. A cache of nuts and seeds in a rock crevice may belong to a chipmunk or white-footed mouse; the same type of sign, with possibly an ear of corn in addition, in a hollow tree or building suggests food of squirrels; and a pile of nuts and seeds in a ground nest probably belongs to a deer mouse. A similar nest containing roots might be that of a meadow vole. In an orchard, a runway leading up to a gnawed apple on the ground indicates the work of a pine mouse. The combination of cut-down trees, sectioned twigs, and wide tooth marks in bark can mean only the work of a beaver.

SYNOPSIS OF ILLINOIS MAMMALS

This section of the Fieldbook furnishes the hiker and naturalist with a concise account of the native and naturalized mammals of Illinois. Keys are provided to aid in identifying the animals, descriptions are given as a means of checking the results obtained by use of the keys, and brief statements on the natural history and distribution are added to assist the user of this book in learning about the mammals.

To make identification of a specimen accurate and yet as easy as possible, each key is divided into two parts. The first part employs only external characters, including the teeth, which may be examined as they are situated in the head, and should be used for identifying the whole specimen or a study skin with a skull.

The second part of each key is based entirely on skull and tooth characters. This part can be used for identifying skulls or incomplete specimens found in the field, in stomach contents of animals, or in owl pellets.

The scientific name of the species precedes the account of each kind of Illinois mammal discussed in the Fieldbook. This scientific name includes two words, first the name of the genus to which the species belongs and second the name of the species; these names are followed by the name of the man who first described the animal. The name of the describer is in parentheses if the animal is not now assigned to the genus in which it was first placed. Below the scientific name is at least one common name. When two or more common names are given, the name on the left side of the page is the one preferred by the authors of this Fieldbook. Each complete account is divided into the following sections: description, life history, signs, and distribution. The description is a summary of the diagnostic characters of the species, including such features as size, color, and dental formula. The life history section includes information on the life cycle, habits, and habitat. The section on signs outlines characteristic identifying features such as nests, scats, food remains, and runways. The section on distribution notes the abundance of the species, its known range in Illinois, the subspecies known to occur in the state, and a brief statement of the known range of the species in North America.

At the end of the accounts of the mammals known to occur in Illinois during historic times, there is a section dealing with species of prehistoric times. Following this are a glossary of terms used in the keys and text and a list of books on mammals.

How to Use the Keys.—In this Fieldbook, a key to the orders of Illinois mammals is given, page 41, and, under each order, a key to the Illinois species of that order. The key to the orders is intended to help you decide which of the keys to the species to use in identifying a specimen.

The keys are made up of pairs of contrasting statements (in abbreviated form) known as key couplets, each couplet preceded by a numeral. In using a key, start at the beginning, read the two contrasting statements of the first couplet, and decide which of the statements best fits the specimen you are trying to identify. What you find at the end of the statement selected will give you a clue as to your next step. If a number is at the end of the statement, follow down the key to the couplet having that number. Repeat the selection process until you have come to a statement having a name at the end of it: the name of an order in the key to the orders and the name of a species in the key to the species.

Some of the couplets contain references to illustrations that will aid you in the identification of your specimen.

If you do not understand terms used in the keys, turn to the glossary for definitions.

Occasionally a specimen will key out readily but be misidentified because a statement in the key has been misinterpreted or because the specimen is abnormal in one or more characters. It is well, therefore, to check your identification by comparing your specimen with the description of the species and the picture of the mammal. It is frequently helpful to consult descriptions of closely related or similar species to gain a knowledge of some of the comparative characters used as a basis for diagnosis. Sometimes the ranges of the species under consideration may help you to decide. If, after following these steps, there is uncertainty as to the identity of a specimen, it should be compared with identified specimens in a reference collection, such as that maintained by the Illinois Natural History Survey, the University of Illinois Natural History Museum, or the Chicago Natural History Museum, or sent to a trained specialist for identification.

Professional mammalogists are usually glad to determine difficult-to-identify specimens sent to them by collectors if the specimens are reasonably well preserved and well packed for shipment, and if accurate locality data accompany them.

Class Mammalia.—Mammals belong to a large group of animals called the Vertebrata, characterized by having in each adult a backbone or a segmented spinal column. The most primitive vertebrates, which include some eel-like forms and the fishes, live in water. Many of the more highly specialized forms of vertebrates, at least in the adult stage, have four feet or limbs instead of fins. Some of the familiar groups of Vertebrata, called classes, are as follows:

Class CHONDRICHTHYES—sharklike fishes

Class OSTEICHTHYES—bony fishes

Class AMPHIBIA—salamanders and frogs

Class REPTILIA—turtles, lizards, crocodiles, and snakes

Class AVES—birds

Class MAMMALIA—mammals

The Mammalia comprise three groups very different in structure and habits: One group, the egg-laying mammals or Monotremata, is confined to the Australasian Region. The second group, the pouched mammals or Marsupialia, is primarily Australian and South American in distribution but is represented in Illinois by a single species, the opossum. The third group, the placental mammals or Eutheria, is the dominant mammal group in most parts of the world and to it belong all the species of the Illinois fauna except the opossum.

Members of the class Mammalia are not only abundant but are also of many different kinds, such as bats, mice, deer, wolves, elephants, and whales. The most closely related kinds are grouped together into *families,* and related families are grouped together into *orders.* The 59 species of wild mammals known to occur in Illinois represent 7 orders and 16 families. Some well-known orders not found in Illinois include the whales and porpoises (order Cetacea), which occur only in the ocean; the sloths and armadillos (order Edentata), which occur in the American tropics and as far north as Kansas; and the manatees and sea cows (order Sirenia), which occur in seas and estuaries in Florida and many other parts of the world.

Checklist of Present Native and Naturalized Mammals of Illinois

Order MARSUPIALIA

Family DIDELPHIDAE—opossums
  Didelphis marsupialis   Opossum

Order INSECTIVORA

Family TALPIDAE—moles
   Scalopus aquaticus   Eastern mole
Family SORICIDAE—shrews
   Sorex cinereus   Masked shrew
   Sorex longirostris   Southeastern shrew
   Microsorex hoyi   Pigmy shrew
   Cryptotis parva   Least shrew
   Blarina brevicauda   Short-tailed shrew

Order CHIROPTERA

Family VESPERTILIONIDAE—bats
   Myotis lucifugus   Little brown bat
   Myotis austroriparius   Southeastern bat
   Myotis grisescens   Gray bat
   Myotis keenii   Keen's bat
   Myotis sodalis   Indiana bat
   Lasionycteris noctivagans   Silver-haired bat
   Pipistrellus subflavus   Eastern pipistrel
   Eptesicus fuscus   Big brown bat
   Lasiurus cinereus   Hoary bat
   Lasiurus borealis   Red bat
   Nycticeius humeralis   Evening bat
   Corynorhinus rafinesquii   Southeastern big-eared bat

Order CARNIVORA

Family PROCYONIDAE—raccoons
   Procyon lotor   Raccoon
Family MUSTELIDAE—weasels, skunks, etc.
   Mustela rixosa   Least weasel
   Mustela frenata   Long-tailed weasel
   Mustela vison   Mink
   Lutra canadensis   River otter
   Mephitis mephitis   Striped skunk
   Taxidea taxus   Badger
Family CANIDAE—wolves, foxes, etc.
   Vulpes fulva   Red fox
   Urocyon cinereoargenteus   Gray fox
   Canis latrans   Coyote
Family FELIDAE—cats
   Lynx rufus   Bobcat

Order RODENTIA

Family Sciuridae—squirrels

Marmota monax	Woodchuck
Citellus tridecemlineatus	Thirteen-lined ground squirrel
Citellus franklinii	Franklin's ground squirrel
Tamias striatus	Eastern chipmunk
Sciurus carolinensis	Eastern gray squirrel
Sciurus niger	Eastern fox squirrel
Glaucomys volans	Southern flying squirrel

Family Geomyidae—pocket gophers

Geomys bursarius	Plains pocket gopher

Family Castoridae—beavers

Castor canadensis	Beaver

Family Cricetidae—native mice and rats

Reithrodontomys megalotis	Western harvest mouse
Peromyscus maniculatus	Deer mouse
Peromyscus leucopus	White-footed mouse
Peromyscus gossypinus	Cotton mouse
Peromyscus nuttalli	Golden mouse
Oryzomys palustris	Rice rat
Neotoma floridana	Eastern wood rat
Synaptomys cooperi	Southern bog lemming
Microtus pennsylvanicus	Meadow vole
Microtus ochrogaster	Prairie vole
Pitymys pinetorum	Pine vole
Ondatra zibethicus	Muskrat

Family Muridae—Old World mice and rats

Rattus rattus	Roof rat
Rattus norvegicus	Norway rat
Mus musculus	House mouse

Family Zapodidae—jumping mice

Zapus hudsonius	Meadow jumping mouse

Order LAGOMORPHA

Family Leporidae—hares, rabbits

Lepus townsendii	White-tailed jackrabbit
Sylvilagus floridanus	Eastern cottontail
Sylvilagus aquaticus	Swamp rabbit

Order ARTIODACTYLA

Family CERVIDAE—deer
  Odocoileus virginianus   White-tailed deer

KEY TO ORDERS

Whole Animals

1. Forelimbs in the form of wings, fig. 39...................
   ....Order CHIROPTERA; bats
   Forelimbs in the form of legs............................ 2
2. Feet each with 1 or 2 large, hard hoofs.................. 3
   Feet each with 4 or 5 clawed toes....................... 4

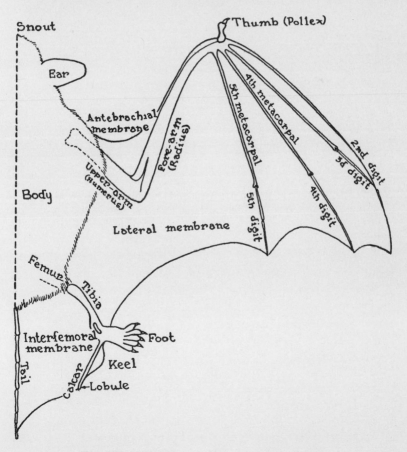

Fig. 39.—Detailed outline of wing of bat.

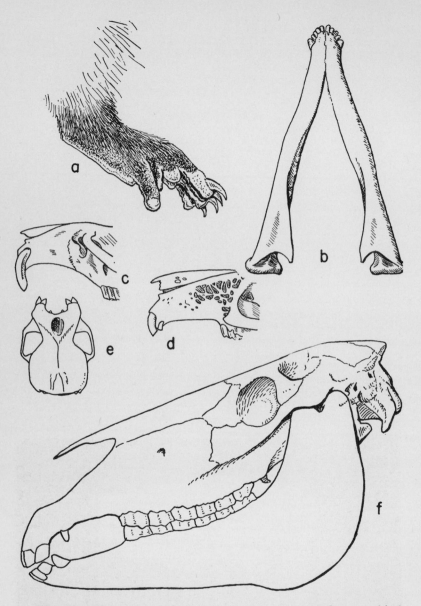

Fig. 40.—Characters referred to in the key to orders: *a,* hind foot of opossum; *b,* lower jaw of opossum, under side; *c,* rostrum of muskrat, side view; *d,* rostrum of cottontail, side view; *e,* skull of bat, top view; *f,* skull of horse, side view.

3. Feet each with only 1 hoof..Order PERISSODACTYLA; horse
   Feet each with 2 hoofs....................................
   ....Order ARTIODACTYLA; deer, cattle, etc.
4. Inner toe of each hind foot thumblike, fig. 40a, widely separated from the other toes; tail long, nearly naked, prehensile, terminal half white ....
   ....Order MARSUPIALIA; opossum
   Inner toes of hind feet not thumblike; tail haired or naked,
   not prehensile, terminal half not white..................   5
5. Canine teeth decidedly longer than adjacent teeth, fig. 117
   ....Order CARNIVORA; flesh eaters
   Canine teeth either much smaller than incisor teeth or absent, figs. 40c, d, 41.....................................   6
6. Canine teeth small, fig. 41; distance between teeth not
   greater than length of a single grinding tooth; eyes and
   ears either minute or absent; snout long, figs. 45–49.....
   ....Order INSECTIVORA; shrews and moles
   Canine teeth absent, fig. 40c, d; gap between incisors and
   grinding teeth equal to or greater than length of entire
   row of grinding teeth; eyes, and usually ears, conspicuous; snout not greatly elongate........................   7
7. Tail a cotton-like tuft; a small peglike tooth immediately
   behind each large upper incisor, fig. 40d...............
   ....Order LAGOMORPHA; rabbits
   Tail elongate, not a cotton-like tuft; no peglike teeth immediately behind the upper incisors....................
   ....Order RODENTIA; rodents

Skulls

1. Upper jaw without incisors..............................
   ......Order ARTIODACTYLA (in part); cattle, deer, etc.
   Upper jaw with incisors................................   2

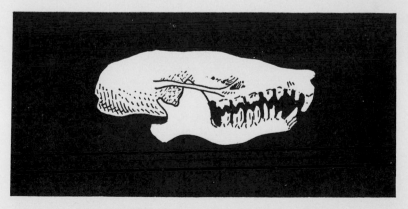

Fig. 41.—Skull of eastern mole, side view.

2. Teeth 50, upper jaw with 26, lower jaw with 24; each posterior angle of lower jaw with an inwardly curving
   process, fig. 40b; nasal bones broadened posteriorly.....
   ....Order MARSUPIALIA; opossum
   Teeth 44 or less, never more than 22 in either jaw; each
   posterior angle of lower jaw without an inwardly curving
   process; nasal bones not broadened posteriorly.......... 3
3. Canine teeth decidedly larger than adjacent teeth, fig. 117.. 4
   Canine teeth either decidedly smaller than incisors or absent, figs. 40c, d, 41................................... 6
4. Anterior end of skull with a wide, U-shaped notch, fig. 40e;
   length of skull less than 20 mm. (about ¾ in.) ....
   ....Order CHIROPTERA; bats
   Anterior end of skull either without a notch or with only
   a small angular or narrow notch, figs. 60, 108; length of
   skull more than 20 mm................................ 5
5. Canine teeth subtriangular in cross section................
   ....Order ARTIODACTYLA (in part); pig
   Canine teeth rounded in cross section....................
   ....Order CARNIVORA; flesh eaters
6. Upper jaw having 6 incisors of approximately equal size,
   fig. 40f; length of skull more than 250 mm. (9¾ in.)....
   ....Order PERISSODACTYLA; horse
   Upper jaw having 2 conspicuously large incisors (in some
   groups followed by minute incisors), fig. 40c, d; length of
   skull less than 150 mm. (5⅞ in.)...................... 7
7. Gap, if present between front teeth and grinding teeth, no
   more than length of a single grinding tooth; canines
   small, fig. 41.....Order INSECTIVORA; moles and shrews
   Gap between front teeth and grinding teeth equal to or
   greater than length of the row of grinding teeth; canines
   absent, figs. 40c, d, 73d................................ 8
8. A pair of peglike teeth immediately behind the 2 large incisors in upper jaw; maxillary region in front of each eye
   socket with bony lattice-work, fig. 40d.................
   ....Order LAGOMORPHA; rabbits
   No peglike teeth behind the 2 large incisors in upper jaw;
   maxillary regions solid bone, fig. 40c...................
   ....Order RODENTIA; rodents

ORDER MARSUPIALIA

Marsupials

The Marsupialia are unique among mammals in that the
female gives birth to minute living young and carries them for
some time, usually in a pouch of skin (marsupium) or in fur
situated on the under side of her body. The marsupials contain
both large and small forms, including the famous kangaroos,
which may attain almost the size of a small horse, and the little

koala pictured on Australian stamps. Marsupials are most abundant in Australia and neighboring islands, but they have many representatives in South and Central America. The range of one species, the opossum, includes much of the United States.

Economic Status.—In Illinois the opossum is trapped for fur and, to some extent, for food. For both food and fur the monetary value is low. The opossum is beneficial in that it eats some injurious insects, but it may at times be harmful to wild birds and cause loss to farmers by eating poultry and eggs.

DIDELPHIS MARSUPIALIS Linnaeus

Opossum   **Possum**

Description.—The opossum, fig. 42, is as large as the average domestic cat. Its face is long and subconical, its ears are naked, its tail is scaly and almost bare, and its feet are decidedly handlike. The female has on her abdomen a woolly pouch in which she carries and nourishes her newborn young. The possum's face is whitish, its eyes are beady black, and its bare ears are shiny black, mottled with pink or white. Its feet are black, and its toes are pink or white. The inner toe of each hind foot is thumblike, fig. 40a.

The hair making up the dense woolly underfur of most opossums is creamy white, with grayish tips, and the long, guard hairs are dark gray or black. This combination gives the body a general grayish appearance. However, in many specimens the underfur is tipped with brown, which, showing through the guard hairs, gives the body a brown instead of a gray appearance. A few individuals are almost entirely black, and others are a very pale gray or nearly white. In the occasional true albino, the general color is off-white, the ears and feet are white, and the eyes are red. Some very light gray individuals may resemble albinos but in these the eyes, ears, and feet have the normal black coloration.

Length measurements: head and body 17–21 inches (430–530 mm.); tail 8½–12½ inches (220–320 mm.); over-all 25½–33½ inches (650–850 mm.); hind foot 2⅜–3⅛ inches (60–80 mm.); ear 2¼–2½ inches (57–66 mm.). Weight (adults, Illinois): 6–12 pounds.

The opossum skull, 100–125 mm. (4–4⅞ inches) long, can be distinguished from the skulls of other Illinois mammals by the

Fig. 42.— Opossums: top, adult; bottom, young opossums about 1 week of age in mother's pouch.

small braincase, the inflected dentary bone, fig. 40*b*, and the large number of teeth (total 50). The sutures between the skull bones tend to remain open, and there is little fusion of bones even in the oldest individuals. Dental formula:* I 5/4, C 1/1, Pm 3/3, M 4/4.

Life History.—The opossum, an inhabitant of woodland, fig. 1, attains sexual maturity at the age of approximately 1 year. The female may have one or, rarely, two litters annually, the first usually in late February and the second in late July. About 13 days after mating, she gives birth to 7 to 20 young, each no larger than a honey bee. The young are nude, sightless, and scarcely more developed than small embryos. By their own strength and instinct they make their way from the birth canal to the mother's pouch. The mother assists them only by making a moist path through the abdominal fur with her tongue and in sitting nearly upright so that the tiny babies will not lose their way and become entangled in dry hair. When they get to the pouch, each one seizes a milk-providing nipple on which it will retain its hold for several weeks. There are only 13 nipples in the pouch and, if the number of young is more than 13, the superfluous number starve.

When the baby opossums are 50 days old, they are about as large as full-grown house mice. By the end of 2 months, their eyes open, and then for the first time they release their hold on the nipples. After 10 more days, they leave their mother's pouch to clamber about her body, fig. 42, but remain nearby. When about 80 days old, they are as large as Norway rats. At this time, they go adventuring on the ground but still return to their mother's pouch for nourishment. If the litter is large, they must nurse in relays. When the young are about 3 months old, the brood gradually breaks up, and each member wanders about until it finds a suitable home in a hollow tree or log, or in a woodchuck or skunk burrow.

When winter comes, the opossum lines its nest thickly with dry leaves or grass and spends the coldest periods dozing in the warmth of a coat of fat gained from its autumn feasts. Only when the nights are not so cold does it come forth to forage, and then it soon hustles back to its den so that its bare ears and tail do not become frostbitten.

---

*An explanation of dental formula is given in the glossary.

This slow, dull, and solitary nocturnal animal eats a variety of food. The chief items of its diet are fruits, insects, some small birds, mice, garbage, and an abundance of scraps from the kills of other mammals. An area of about 12 acres of good woodland that is not heavily grazed and that has plenty of food and water is probably sufficient to support one opossum.

Abundance of food and suitable habitats and production of a large number of young would soon result in a tremendous num-

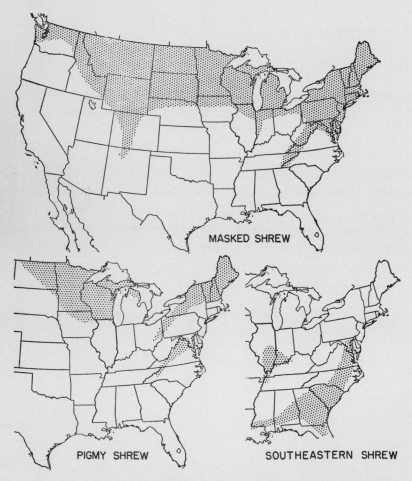

Fig. 43.—Known distribution, in the United States, of insectivores with a limited range in Illinois.

ber of opossums if it were not for their enemies. Foxes, dogs, coyotes, possibly minks, and large owls pick them off one by one. Although opossums are rather easily caught, they have a unique method of defense. If surprised away from the comparative safety of trees, they may feign death so that a not-too-hungry enemy may leave them for dead without trying to eat them. Sometimes, however, they are badly mauled before being left alone. Considerable numbers are killed by motor vehicles on highways. Several hundred thousand possums are caught each year by trappers or fur hunters in Illinois, even though possum pelts are worth very little.

Signs.—Possum tracks, fig. 17, easy to detect in mud or snow, are most apt to be found at the edges of ponds and rivers near woodlands. There is no mistaking them. The print of each front foot shows five fingers spread wide apart, and that of each hind foot shows the large toe thrust out at a right angle to the other toes, which are rather close together.

Ordinarily possum droppings are in the form of irregular masses about 2 to 2½ inches long; usually they contain large amounts of fruit hulls and seeds and often some fur and feathers. The fur and feathers are more than likely from various creatures that have been killed by some predator other than the possum.

Distribution.—The opossum is common in Illinois. The subspecies in this state is *Didelphis marsupialis virginiana* Kerr. The species has a range that includes an extensive area: all of the eastern United States south of a line drawn from southern New England to southern Minnesota to western Texas; also much of western California. The range extends through Mexico and into South America.

ORDER INSECTIVORA
Moles and Shrews

The Insectivora are the most primitive of the known living placental mammals; all Illinois mammals except the opossum are placental. Several families of insectivores are recognized. Two of these, the Talpidae or moles and the Soricidae or shrews, are burrowing, thick-furred animals comparable in size to rats and to small mice, respectively. These two families, widespread in the Northern Hemisphere, are represented in

Illinois. The known distribution, in the United States, of three shrews having a limited range in Illinois is shown in fig. 43.

Economic Status.—All members of the order are beneficial to some extent, because they feed on soil-inhabiting insects, many of which are injurious to crop plants. Moles and shrews are of no commercial value for their pelts. Moles often become nuisances in gardens and lawns by raising the soil above their subsurface runways.

KEY TO SPECIES

Whole Animals

1. Front feet more than twice as wide as hind feet; over-all
   length of animal more than 150 mm. (5⅞ in.)..........
   ....Talpidae (moles)   2
   Front feet approximately same width as hind feet; over-all
   length of animal less than 130 mm. (5⅛ in.)...........
   ....Soricidae (shrews)   3
2. Snout naked; eyes not visible; tail length less than 40 mm.
   (1⅝ in.).................eastern mole, *Scalopus aquaticus*
   Snout with fleshy, finger-like tentacles; eyes visible; tail
   length more than 50 mm. (2 in.)......................
   ....star-nosed mole, *Condylura cristata*
3. Tail length less than half of head+body length; ears hidden in fur........   4

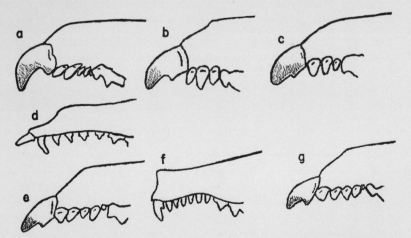

Fig. 44.—Rostrums or upper jaws of insectivores, side view: *a,* short-tailed shrew; *b,* least shrew; *c,* pigmy shrew; *d,* star-nosed mole; *e,* southeastern shrew; *f,* eastern mole; *g,* masked shrew.

Tail length more than half of head+body length; ears not
hidden in fur... 5
4. Over-all length of animal less than 88 mm. (3½ in.) ; fur
gray-brown in color............least shrew, *Cryptotis parva*
Over-all length of animal more than 88 mm.; fur blackish
in color............short-tailed shrew, *Blarina brevicauda*
5.* Upper jaw, in lateral view, with 3 unicuspids visible on
each side, fig. 44c...........pigmy shrew, *Microsorex hoyi*
Upper jaw, in lateral view, with 4 or more unicuspids visible on each side, fig. 44e............................... 6
6.* Third unicuspid smaller than fourth, fig. 44e ....
...................southeastern shrew, *Sorex longirostris*
Third and fourth unicuspids subequal or third larger than
fourth, fig. 44g.............masked shrew, *Sorex cinereus*

Skulls

1. Zygomatic arches present, fig. 41; teeth not tipped with red
or brown.................... ....Talpidae (moles) 2
Zygomatic arches absent; front teeth tipped with red or
brown, fig. 44a–c, e, g...............Soricidae (shrews) 3
2. Teeth in upper jaw 20, in lower jaw 16; upper incisors
slightly recurved, fig. 44f...eastern mole, *Scalopus aquaticus*
Teeth in upper jaw 22, in lower jaw 22; upper incisors
projecting forward, fig. 44d..........................
...................star-nosed mole, *Condylura cristata*
3. Teeth in upper jaw 18 ....least shrew, *Cryptotis parva*
Teeth in upper jaw 20 4
4. Width of skull more than 11 mm. (⅜ in.) ....
...................short-tailed shrew, *Blarina brevicauda*
Width of skull less than 11 mm...................... 5
5. Upper jaw, in lateral view, with 3 unicuspids visible on
each side, fig. 44c............pigmy shrew, *Microsorex hoyi*
Upper jaw, in lateral view, with 4 or more unicuspids visible on each side, fig. 44e, g........................... 6
6. Third unicuspid smaller than fourth, fig. 44e...........
...................southeastern shrew, *Sorex longirostris*
Third and fourth unicuspids subequal or third larger than
fourth, fig. 44g.............masked shrew, *Sorex cinereus*

SCALOPUS AQUATICUS (Linnaeus)
Eastern Mole

Description.—The eastern mole, fig. 45, has a rather thick
body, plushlike fur, short forelimbs, broadly rounded hands
with outwardly turned palms, pointed snout, short tail, no external ears, and no visible eyes.

---

*The characters in this couplet are seen best in a cleaned skull, and only with
difficulty in a whole animal.

The eyes, which are completely covered with skin, may serve the animal only to distinguish between light and dark. The elongate nose is a sensitive probe used in seeking out earthworms, grubs, and other food. The fur appears usually as a dark or slate gray, but in some lights and at certain angles it appears as light gray and, in most individuals, has a slight iridescence. The summer pelage, acquired after a spring molt, is slightly paler than the winter pelage. The fur of the under

Fig. 45.—Eastern mole.

parts is nearly the same color as that of the sides and back, except that on some individuals bright orange or white patches appear on the abdomen or on the chest.

Length measurements: head and body 5½–6½ inches (142–167 mm.); tail 1⅛–1½ inches (28–38 mm.); over-all 6¾–8 inches (170–205 mm.); hind foot about 1 inch (23–25 mm.). Weight (mature male): about ¼ pound (90–125 gm.).

The skull is cone shaped, pointed in front and broad at the back; it is 28 mm. (about 1⅛ inches) long. Zygomatic arches are present. All of the bones of the skull fuse together early in the development of the animal, and the cranium appears to consist of but one large bone. All of the teeth have sharp cusps designed for cutting worms and other items of food, figs. 41, 44f. Dental formula: I 3/2, C 1/0, Pm 3/3, M 3/3.

Life History.—The eastern mole lives in runs, burrows, or tunnels near the surface of the earth, fig. 1. These are the result of the animal's "swimming" through the soil in search of food. In the "swimming" motion a breast stroke is employed. The mole's spadelike hands are brought forward alongside its snout. Then its hands are thrust outward, slightly upward, and backward, pushing the soil aside and pulling the animal forward. The soil above the animal arches and finally cracks, leaving a humped and broken trail, fig. 6. Along such a run the mole searches for food. Rarely, an adventuresome or restless individual will leave its run or burrow for a brief foray above ground. Only on such an occasion does the mole leave open an entrance to its burrow. When the mole is at home, every entranceway is plugged. Meadow mice, pine mice, and shrews commonly dig into and use mole runways.

Although the eastern mole makes unsightly ridges and hills or mounds in lawns or gardens, it has the best of intentions, for usually in this activity it is not eating roots or bulbs but foraging for worms and insects. Examination of the stomachs of eastern moles has demonstrated that the bulk of the diet consists of earthworms, white grubs, and other arthropods. In the stomachs of 56 moles from central Illinois, 62 per cent of the food was insects, 26 per cent earthworms, 11 per cent plant material, and 1 per cent spiders, hair, and other items (West 1910). Eight per cent of the plant material was corn, found in the stomachs of 11 of the 56 moles. In the stomachs of these same animals there were also cutworms, wireworms, white grubs, webworms, and ants. In a mole stomach collected recently in Champaign County, most of the food consisted of garden peas.

The eastern mole is active throughout the year. In spring and summer it uses runways or tunnels near the surface and extends them chiefly when soil conditions are most favorable— usually following rains. In the fall it devotes much of its time to making runs or burrows deep in the soil. With the onset of winter, it uses both new and old runs below the frost line where insects can still be found.

The female of the eastern mole ordinarily has only one litter of four young each year. In April or May, the young are born, naked and helpless, in a crudely constructed nest in a deep runway. For about a month, they remain in the nest. By the age of 3 months they are nearly as large as the parents.

This animal has few enemies and leads a relatively safe life in its underground burrows. To some wild animals the mole is probably distasteful because of a secretion from its skin glands. Cats are known to catch moles but to refuse to eat them.

Signs.—An unweathered mole hill or mound, fig. 6, is so different from other animal diggings that it is not easily mistaken. It resembles the mound made by a pocket gopher, fig. 7, in that no hole shows if the animal is in its burrow; however, much more of the earth of the mole hill is in larger lumps or clods. The dirt plugging the hole is pushed out by each new load thrust up under it until a steep-sided pile, usually less than a foot across and about 5 or 6 inches deep, has been formed.

During the summer, freshly made hills are relatively uncommon, and the presence of moles is usually detected by ridges, 5 to 8 inches in width, of broken surface soil or torn sod.

Distribution.—The eastern mole is common over most of Illinois but uncommon in the extreme northeastern counties. The subspecies in Illinois is *Scalopus aquaticus machrinus* (Rafinesque), sometimes called the prairie mole. The species occurs from southern New England to central Minnesota and northeastern Colorado and south to northern Mexico and southern Florida.

CONDYLURA CRISTATA (Linnaeus)
Star-Nosed Mole

This species has been reported from Illinois on the basis of sight records, but to date no specimens of it from this state have been captured and preserved. It seems best, therefore, to omit the name of this species from the list of Illinois mammals until proof of its occurrence here is established. The star-nosed mole is northeastern in distribution; the records nearest Illinois are from Wisconsin, Michigan, and eastern Indiana. Specimens may be identified by the characters given in the key to insectivores.

SOREX CINEREUS Kerr
Masked Shrew

Description.—The masked shrew, fig. 46, is frequently mistaken for a young mouse but, unlike the mouse, it has a fine, velvety fur, sharp conical muzzle, minute eyes, and pincer-like teeth that are tipped with red. It has a relatively long tail, which reaches nearly to the ears when laid forward along the back. In winter this animal is brown on the upper parts, smoky gray on the under parts; in summer it is a slightly darker shade. The eyes and ears are masked or covered by short hair; hence the common name of masked shrew.

Length measurements: head and body 1⅞–2½ inches (47–63 mm.); tail 1¼–1¾ inches (33–46 mm.); over-all 3⅛–4¼ inches (80–109 mm.); hind foot about ½ inch (10–13 mm.). Weight: about ⅛ ounce (4 gm.).

The skull is slender, tear-drop shaped, and 16.5 mm. (about ⅝ inch) long. The teeth are numerous (32), small, and sharp pointed. On each side of the upper jaw there is one large front incisor which is notched and projected forward. Behind this tooth are five small teeth, called unicuspids (one-cusped teeth). The fourth unicuspid is about the same size as or smaller than the third unicuspid, and the fifth unicuspid is so small that it may be overlooked, fig. 44g. Dental formula: I 3/1, C 1/1, Pm 3/1, M 3/3.

The masked shrew may be confused with the southeastern shrew. The two are best distinguished by differences in the unicuspids. They differ in distribution; the masked shrew has never been taken in southern Illinois, and there are no authentic records of the southeastern shrew from northern Illinois.

The masked shrew may also be confused with the pigmy shrew, but generally the masked shrew is larger, has a longer

Fig. 46.—Masked shrew.

tail and longer hind feet, and has five instead of four unicuspids (three visible in lateral view) on each side of the upper jaw.

Life History.—Little is known about the masked shrew in Illinois. Elsewhere an adult female of this species produces about five young in April. Her ball-shaped nest, 4 or 5 inches in diameter and lined with fine grass and rootlets, is hidden in a cavity of an old log or stump.

The masked shrew is generally found under a dense growth of weeds or in woods, fig. 1. This tiny mammal, like all other shrews, is a scurrying, vibrating mite of energy, driving narrow tunnels in leaf mold and darting swiftly about in search of food day or night, summer or winter. It is a prodigious eater. In a single day a shrew may eat one, two, or even three times as much as its own weight in food. Its diet consists chiefly of insects, snails, and worms.

This shrew, like others in Illinois, possesses scent glands that secrete a musky fluid which renders the animal undesirable as food for some predators and may account for the fact that cats sometimes leave shrews uneaten after catching and killing them.

Signs.—Curious parallel tracks in snow (tracks similar to but smaller than those of the short-tailed shrew), and often accompanied by a tail mark, may indicate the presence of the masked shrew. This animal may make small ridged runways in the snow similar to those of the mole in soil, but the shrew runways are only about ¾ inch across as seen from the surface. The burrow the masked shrew makes in snow, and probably also in loose soil, has an inside diameter of about half an inch.

Distribution.—The masked shrew is uncommon and in Illinois is restricted to the northern fourth of the state, fig. 43. The subspecies occurring in this state is *Sorex cinereus lesueurii* (Duvernoy). The range of this species includes most of Canada and much of the northern United States, with southern extensions in the Rocky Mountains to northern New Mexico, in the Great Lakes region to the Wabash River valley, and in the Appalachian Mountains to western North Carolina.

SOREX LONGIROSTRIS Bachman

Southeastern Shrew   Bachman's Shrew

Description.—The southeastern shrew, fig. 47, is of about the same size and proportions as the masked shrew and can be distinguished with certainty from the latter only by skull characters. The pelage is reddish brown.

Length measurements: head and body about 2⅛ inches (55 mm.); tail about 1 inch (27 mm.); over-all about 3¼ inches (82 mm.); hind foot ⅜ inch (9 mm.).

The skull is shorter than that of the masked shrew but it is similar in shape. The molar teeth are larger than those of the masked shrew; the fourth unicuspid on each side of the upper jaw is larger than the third, fig. 44e. Dental formula: I 3/1, C 1/1, Pm 3/1, M 3/3.

The southeastern shrew can be distinguished from the pigmy shrew in having five rather than three readily discernible unicuspids on each side of the upper jaw.

Life History.—The southeastern shrew is one of the least known mammals in Illinois, for less than a dozen specimens have been collected in the state. One specimen, taken at Fox Ridge State Park, Coles County, was found beneath a log where a pond had dried up in a brushy, sparsely wooded portion of the park. Another individual from the same locality was found in a wooded ravine.

Signs.—Presumably the signs of the southeastern shrew are identical with those of the related masked shrew.

Distribution.—The rare southeastern shrew is known in Illinois only from Alexander, Coles, Fayette, and Johnson counties. Specimens taken in this state are presumed to belong to the subspecies *Sorex longirostris longirostris* Bachman. The

Fig. 47.—Southeastern shrew.

species has a range that appears to be discontinuous, fig. 43. One part of the range apparently extends from Maryland to southwestern Mississippi and northern Florida; another part includes areas in Illinois, Indiana, Kentucky, and Tennessee.

MICROSOREX HOYI (Baird)
Pigmy Shrew

Description.—The pigmy shrew is probably the smallest American mammal. An adult weighs only 2 to 3 grams (less than ⅛ ounce); it would take 8 of these shrews to equal the weight of a white-footed mouse, 400 to equal the weight of a fox squirrel, and several thousand to equal the weight of a bobcat. This shrew is a uniform light brown on the upper parts and slightly paler brown on the lower parts.

Length measurements (based on one Illinois specimen): head and body 1⅞ inches (48 mm.); tail 1⅛ inches (29 mm.); overall 3 inches (77 mm.); hind foot ⅜ inch (9.5 mm.).

The skull is similar in many details to that of the southeastern shrew and that of the masked shrew, but it is smaller; it has only three readily discernible unicuspids on each side of the upper jaw, fig. 44c; the first and fifth are minute. Dental formula: I 3/1, C 1/1, Pm 3/1, M 3/3.

Life History.—Little is known about the pigmy shrew, since it is very rare. The only known Illinois specimen was taken in the middle of the winter of 1949 (Sanborn & Tibbitts 1949) in a garage in Cook County. A masked shrew was found with it, and possibly both shrews were forced there because of a heavy coating of ice out-of-doors.

Apparently the pigmy shrew inhabits dry woodlands, thickets, and grassy clearings, where it feeds largely on insects.

Signs.—Inasmuch as the pigmy shrew is similar to the two preceding species of *Sorex* in body form and habits, its signs are probably also similar.

Distribution.—The only specimen of the pigmy shrew recorded for Illinois is from Palatine, in Cook County. It belongs to the subspecies *Microsorex hoyi hoyi* (Baird). The range of the species includes most of Canada and Alaska except the West Coast. It extends into the United States as far as northern Illinois, eastern Ohio, and down the Appalachians to North Carolina, fig. 43.

BLARINA BREVICAUDA (Say)
Short-Tailed Shrew

Description.—At first glimpse, a short-tailed shrew, fig. 48, scurrying along in leaf litter may be mistaken for a mouse or a young mole. Close examination will show that this shrew has some features of both the mouse and the mole. It is sometimes called the mole shrew. Unlike the mole, the short-tailed shrew has eyes that are functional, although small, and front feet that are not broad and spadelike. This shrew is about the size of a

Fig. 48.—Short-tailed shrew.

mouse but differs from it in having a plush, velvety fur, fig. 4, sharp-pointed nose, short tail, and a seeming absence of ears, which are hidden in the fur.

Length measurements: head and body about 3–4 inches (75–101 mm.); tail ¾–1⅛ inches (20–29 mm.); over-all 3¾–5⅛ inches (95–130 mm.); hind foot about ⅝ inch (14–17 mm.).

The skull ranges in length from 21.0 to 24.5 mm. (less than 1 inch) and in width from 11.3 to 12.9 mm. (about ½ inch). It has no zygomatic arches. Part of the upper jaw is shown in fig. 44a. Dental formula: I 3/1, C 1/1, Pm 3/1, M 3/3.

The short-tailed shrew can be distinguished from all other Illinois mammals by a combination of characters: mouse-size body; short tail; dark, nearly black, velvety fur; sharp-pointed nose; pin-point eyes; relatively large skull (nearly mouse-skull size) without zygomatic arches.

Life History.—The short-tailed shrew is an animal principally of forest floors, forest edges, meadows near woods, or swampy, brushy habitats. Usually it builds its own burrow or tunnel through leaves or humus. A log may serve as a roof for its burrow. Frequently, when a log in the forest has been overturned or broken open, a furrow-like runway of the short-tailed shrew is revealed. Burrows of this kind, about 1¼ inches in diameter, may be so numerous as to form a network in the forest floor. A man walking through a forest may find that at nearly every step his heel sinks into one of these burrows an inch or two below the surface, fig. 1. Commonly there are 10 to 20 short-tailed shrews per acre in forests and fewer per acre in grassy, less wooded areas. This shrew is regarded as the most abundant small mammal in many wooded areas of Illinois.

Its nest of grass, leaves, or hair, or a combination of these materials, may be hidden beneath a large log or in a subterranean burrow. This nest is slightly oval in shape and 4¾ to nearly 6 inches (120–150 mm.) in diameter. The number of litters a female may have in a year is not known; probably there are two or three litters of five to eight young each. The gestation period is 21 to 22 days. At birth the young are helpless, pink and wrinkled, and each is smaller than a honey bee.

The short-tailed shrew is a ravenous feeder, consuming the equivalent of its weight in food in a single day. Its food consists principally of invertebrates, such as earthworms, snails, and insects. Short-tailed shrews in captivity have been observed attacking and killing mice, but no one has reported seeing a shrew do this in the wild. The production of a poisonous substance by the submaxillary glands may aid this shrew in overcoming its victim. In wintertime it caches live snails in its burrows below ground and from this "pantry" it can draw food as needed.

Shrews of this species occur in the diet of owls and snakes. Because uneaten carcasses of shrews are rather frequently found, it is thought that some predators kill these animals and then find them unpalatable.

Short-tailed shrews may be kept in captivity, but only one shrew should be placed in a container. If two are put together, one will soon kill and devour the other. A glass aquarium tank, with wood shavings in the bottom and a half-pint glass jar placed on its side for a nest chamber, makes an ideal cage. Water should be available at all times, and food should be offered in liberal quantity. An ideal food, used with success by Dr. Oliver P. Pearson of the University of California, who has raised many of these shrews, consists of about equal parts of dry dog meal and hamburger, or horse meat, mixed to hamburger consistency by the addition of water. This diet can be supplemented with worms, beetles, and grasshoppers. With proper care, short-tailed shrews may live in captivity to be nearly 3 years old, but, in the wild, their life-expectancy probably never exceeds 1 year.

Signs.—The usual gait of the short-tailed shrew in soft snow is a walk, and the resulting tracks consist of unpaired footprints evenly placed along the trail, fig. 33, at about 1-inch intervals, with the tail mark between them curved gently from side to side as a result of the swaying of the short-legged body. A running shrew leaves paired footprints and an interrupted tail mark; the distance between each set of paired prints then is about 5 inches.

Droppings of the short-tailed shrew are greenish black when fresh, slightly brownish when dry, spindle shaped, about a third of an inch long, and coiled in various ways.

Distribution.—The short-tailed shrew is a common species throughout Illinois. The subspecies *Blarina brevicauda brevicauda* (Say), with larger individuals, occurs in the northern part of the state, and the subspecies *B. b. carolinensis* (Bachman), with smaller individuals, occurs in the southern part. The area of intergradation between these two subspecies is poorly known. The range of the species embraces roughly the southeastern fourth of North America; it extends northward into southern Canada and westward to about the 100th meridian.

CRYPTOTIS PARVA (Say)

Least Shrew   Old-Field Shrew

Description.—The least shrew, fig. 49, is a gray-brown miniature of the short-tailed shrew. It can be distinguished from the masked, southeastern, and pigmy shrews by its shorter tail and usually by its more grayish color and more effectively concealed ears. It differs from the short-tailed shrew by its smaller size, fig. 4, and grayish brown rather than blackish color.

Length measurements: head and body about 2⅛–2½ inches (54–63 mm.); tail ½–¾ inch (11–17 mm.); over-all about 2½–3⅛ inches (65–80 mm.); hind foot ⅜ inch (9–11 mm.).

The skull length is 16.0–17.0 mm. (about ⅝ inch); width, 7.5–8.2 mm. (slightly more than ¼ inch). The least shrew has only two premolars on each side in the upper jaw, whereas all other Illinois shrews have three. Part of the upper jaw is shown in fig. 44b. Dental formula: I 3/1, C 1/1, Pm 2/1, M 3/3.

Life History.—The least shrew often is found in old, weedy fields, fig. 2, and thus it is sometimes called the old-field shrew. It is frequently common in bluegrass meadows, occupying runways of meadow mice. It may occur also in brushy, weedy, or

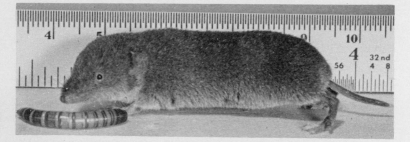

Fig. 49.—Least shrew.

marshy situations near woods but rarely, if ever, in forests. It probably occupies burrows and runways made by various other mammals, but some surface runs may represent paths made and used exclusively by the least shrew.

Except for *Blarina brevicauda,* the least shrew is the commonest shrew in Illinois. In some fields, least shrews may be as abundant as 10 or 15 per acre.

It is not known when the least shrew breeds in Illinois, when it brings forth young, or how many young are in each litter. According to limited observations on this species in other states, a litter may consist of five or six young. The combined weights of these young while still nursing may be several times that of the mother.

Like other shrews, the least shrew has an insatiable appetite. An individual kept in captivity ate seven migratory grasshoppers in 30 hours. It killed each one by biting the head, and then it proceeded to eat the insect head first, discarding the wings and hind legs as it came to these structures. The captive animal clambered slowly about the hand of its human captor, occasionally trying to bite, but it was never successful in breaking the skin. In its natural habitat, this shrew probably preys on spiders, snails, and worms, as well as on insects.

Signs.—Tracks of the least or old-field shrew are similar in most respects to those of the short-tailed shrew, but they are no more than half as large. The tail is too short to leave a drag mark in soft snow. In snow or soil, the burrow of this shrew is only about ¾ inch in diameter.

Distribution.—The least shrew is probably state-wide in occurrence, but it seems to be rare in the northern quarter of the state. Two subspecies occur in Illinois, *Cryptotis parva parva* (Say) occupying the northern two-thirds and *C. p. harlani* (Duvernoy) the southern third. The species ranges from Connecticut to northeastern Colorado and southward into Mexico and Florida.

ORDER CHIROPTERA
Bats

Bats are the only mammals that truly fly. They have expansive membranous wings formed of thin skin that extends from the sides of their bodies out over frames composed of the conspicuously long bones of their forearms and fingers down to their hind legs, fig. 39; in many species, this thin skin encompasses the tails also. Usually bats fly only at night or in the twilight of evening or of early morning. They are extremely agile in flight and, although they have poor eyesight, they are adept at avoiding obstacles and successful in catching insects on the wing. They guide themselves in the dark by means of a unique sonar system. They emit from their throats squeaks and supersonic vibrations, inaudible to man, and perceive and localize the reflected sound waves or echoes through highly developed mechanisms of their ears.

Bats are distributed over the whole world and are of many species. One famous bat is the vampire, which is of medium size and is found in the American tropics. It punctures or shaves the skin of large mammals (occasionally man) and laps up the blood as food. The largest bats are the fruit bats or flying foxes of the South Pacific tropics, some attaining a body length of a foot and a wing span of 5 feet. The 12 species of bats in Illinois, all belonging to the family Vespertilionidae, are harmless and small, the largest being only 5½ inches from nose to tip of tail and having a wing span of about 14 inches.

The species of bats in Illinois can be divided into two general groups. Individuals in one group tend to be solitary (live alone); to roost in trees (hang from branches with their heads down or hide under loose bark); and to be migratory (move south, probably out of Illinois, late in the fall and north in the spring). This group includes the red, the hoary, and the silver-haired bats.

Individuals in the second group tend to be gregarious (live together in colonies); to hibernate in caves or abandoned mines in winter and roost in buildings, caves, or hollow trees the remainder of the year; and to be nonmigratory (except in one species).

Mating among Illinois bats typically occurs late in the fall, ovulation and fertilization the following spring. The young are born in late spring or early summer. At this time the sexes usually segregate; some colonies consist entirely of males and others of females with young. The number of young produced annually by a female is usually one or two; the number varies with the species involved. The bare, much wrinkled, and blind young have proportionately small hands, or wings, and modified bladelike or forked and recurved milk teeth. These forked teeth presumably aid the young in clinging tightly to the body of the mother, inasmuch as she carries them until they are too much of a burden or are old enough to fly. Sometimes the combined weight of two young exceeds that of the mother. Development is extremely rapid, some young being able to fly at the age of 3 weeks.

Signs.—Bats leave signs of their presence in the form of droppings, or guano, and the remains of partially eaten insects. Because such signs are the same for most species of bats, a brief description of them is summarized here, rather than in the description for each species. Bats of the first group mentioned above, sometimes referred to as the tree bats, seldom leave sufficiently large accumulations of guano and insect remains to be easily recognized as bat signs. Bats of the second group leave deposits of droppings and insect remains that may be several feet deep under the roosting sites. Caves long inhabited by summer colo-

nies of bats are easily recognized by the mounds of guano and
sometimes by the presence of dark stains on the ceilings where
clusters of bats have been suspended. The known distribution,
in the United States, of five species of bats having a limited
range in Illinois is shown in fig. 50.

Economic Status.—Bats in Illinois are undoubtedly beneficial because they feed primarily on insects and aid in reducing
the numbers of some important pests. Colonies of bats in build-

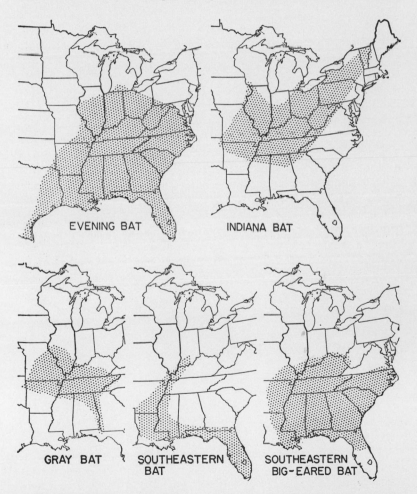

Fig. 50.—Known distribution, in the United States, of bats with
a restricted range in Illinois.

ings may be objectionable because the droppings that accumulate produce an odor that generally is regarded as unpleasant.

KEY TO SPECIES

Whole Animals

Bats are rather difficult to identify; often a fresh specimen can be more readily identified than a study skin or preserved specimen. The following key to the whole animals is designed for living or freshly killed bats and for study skins in which detailed measurements accompany the skin and skull. In each instance, the ear length, forearm length, or foot length is that of a fresh specimen.

1. Ear more than 30 mm. (1⅛ in.) long; glandular outgrowth or lump on side of muzzle..............................
.... southeastern big-eared bat, *Corynorhinus rafinesquii*
Ear less than 23 mm. (⅞ in.) long; no prominent lump on side of muzzle 2
2. Upper surface of tail membrane completely furred, fig. 56; under side of wing with a patch of fur on distal part of forearm 3
Upper surface of tail membrane entirely bare, fig. 55, or at least posterior third bare; under side of wing naked ... 4
3. Fur brick or rusty red, dusted with white; forearm less than 44 mm. (1¾ in.) long; over-all length less than 115 mm. (4½ in.) .... red bat, *Lasiurus borealis*
Fur a mixture of bright buff, yellows, and deep amber, heavily frosted with white in waves over the back; forearm more than 44 mm. long; over-all length more than 130 mm. (5 in.)................hoary bat, *Lasiurus cinereus*
4. Fur black or blackish brown, frosted with white; premolars on each side of jaws 2 above, 3 below....................
.... silver-haired bat, *Lasionycteris noctivagans*
Fur neither black nor frosted with white; premolars on each side of jaw not 2 above, 3 below 5
5. Over-all length usually 105 mm. (4⅛ in.) or more; forearm length usually 45 mm. (1¾ in.) or more ....
.... big brown bat, *Eptesicus fuscus*
Over-all length usually less than 105 mm.; forearm length less than 45 mm. 6
6. Fur on back with hairs not darkened at bases; forearm length usually 40 mm. (about 1½ in.) or more..........
.............................gray bat, *Myotis grisescens*
Fur on back with hairs darkened at bases; forearm length usually less than 40 mm.................................... 7
7. Fur on back with hairs distinctly tricolored; forearm and fingers red; upper jaw with 2 pairs of premolars ..
.....................eastern pipistrel, *Pipistrellus subflavus*
Fur on back with hairs not distinctly tricolored; forearm

and fingers brown or black; upper jaw with 1 or 3 pairs of premolars.. 8

8. Upper jaw with 1 pair of incisors; tragus curved and less than 4 mm. (about ⅛ in.) from notch to tip..........
........................evening bat, *Nycticeius humeralis*
Upper jaw with 2 pairs of incisors; tragus straight and more than 4 mm. from notch to tip 9

9. Ear large, extending 3 or 4 mm. (about ⅛ in.) beyond end of snout when laid forward; tragus slender and pointed, fig. 51*b*........................Keen's bat, *Myotis keenii*
Ear not extending 3 or 4 mm. beyond end of snout when laid forward; tragus broadly rounded, fig. 51*a* 10

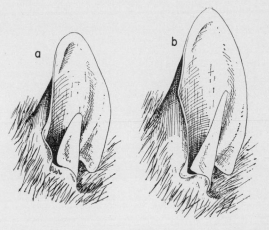

Fig. 51.—Ears of two bats: *a,* little brown bat; *b,* Keen's bat.

10. Fur on under side of body white or pale gray; fur on upper side of body dense, woolly, and brown, with a mole-gray cast; nose flesh-colored in living bat............... ...
....................southeastern bat, *Myotis austroriparius*
Fur on under side of body buffy gray or pinkish gray; fur on upper side of body long, straight, brown, but without a mole-gray cast; nose dark colored in living bat........ 11

11. Forearm less than 35 mm. (1⅜ in.) long; foot not more than 8 mm. (less than ⅜ in.) long ....
.................small-footed brown bat, *Myotis subulatus**
Forearm more than 35 mm. long; foot more than 8 mm. long 12

12. Fur on upper parts of body with coppery or bronzy tipped hairs; on under parts buffy gray; calcar lacking a definite keel; foot usually 10 mm. long..................
.. .. little brown bat, *Myotis lucifugus*

_____
*This species may occur in Illinois, but there are no official records of it.

Fur on upper parts of body without bronzy tipped hairs;
on under parts pinkish gray; calcar with a small but definite keel, fig. 39; foot usually 9 mm. long...............
...........................Indiana bat, *Myotis sodalis*

Skulls

1. Upper jaw with 1 pair of incisors...................... 2
   Upper jaw with 2 pairs of incisors...................... 3
2. Upper jaw with 12 teeth ...evening bat, *Nycticeius humeralis*
   Upper jaw with 14 teeth.....................*Lasiurus* spp.
3. Upper jaw with 14 teeth ....big brown bat, *Eptesicus fuscus*
   Upper jaw with 16 or more teeth...................... 4
4. Lower jaw with 18 teeth (2 pairs of premolars).........
   ....eastern pipistrel, *Pipistrellus subflavus*
   Lower jaw with 20 teeth (3 pairs of premolars) .   5
5. Upper jaw with 18 teeth (3 pairs of premolars)...*Myotis* spp.
   Upper jaw with 16 teeth (2 pairs of premolars).......... 6
6. Rostrum (dorsal view) almost as wide as braincase......
   ....silver-haired bat, *Lasionycteris noctivagans*
   Rostrum (dorsal view) about half as wide as braincase....
   ....southeastern big-eared bat, *Corynorhinus rafinesquii*

MYOTIS LUCIFUGUS (Le Conte)

Little Brown Bat

Description.—The little brown bat, fig. 52, is of medium
size but, like all other bats, it appears to be much larger than
it actually is. Although its wingspread may be nearly 10 inches,
the animal weighs only a quarter of an ounce, and the head
and body are only about 2 inches long.

The upper parts of the body have olive-brown or yellowish
brown fur with a bronzy sheen, and the under parts have gray
fur washed with buff. The ears, wings, and tail membrane are
dark brown, nearly black, and are almost free of hair.

Length measurements: head and body 1¾–2 inches (45–52
mm.); tail 1¼–2 inches (30–50 mm.); over-all 3¼–3¾ inches
(82–95 mm.); hind foot ⅜ inch (10–11 mm.); ear from notch
⅝ inch (14–16 mm.).

The skull is small; it has a short rostrum and an over-all
length of 14.6–15.1 mm. (about ⅝ inch). The braincase is no
larger than a dried pea. The incisor teeth are so small as to be
hardly visible. Dental formula: I 2/3, C 1/1, Pm 3/3, M 3/3.

Four closely related species—gray bat, Indiana bat, Keen's
bat, and southeastern bat—may occur in the same roosting

place with the little brown bat. The little brown bat can be
distinguished from Keen's bat by smaller ears, which do not
reach beyond the nose when laid forward, and by tragi, fig. 51a,
that are each less than 6.5 mm. long. It differs from the gray
bat by having forearms that are each less than 42 mm. long. It

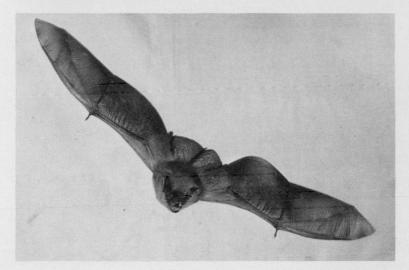

Fig. 52.—Little brown bat.

differs from the Indiana bat by having a bronzy sheen on the
back, and from the southeastern bat by its buffy rather than
nearly white under parts.

The little brown bat differs from the eastern pipistrel by its
brown rather than yellow color and from the evening bat by its
lighter brown color. It also has less blunt tragi than either of
these species.

Life History.—The little brown bat is present in Illinois
throughout the year. During the warm months it congregates
in large numbers in attics and steeples; during cold weather it
hibernates in suitable caves and mines in even larger numbers.
In these hibernating chambers, where the temperature is constantly cool but above freezing, individuals of this species may
hang with their heads down, singly or in clusters of 20 to 100
or more, or they may wedge their bodies into cracks. Here they
become dormant and in a state of deep hibernation; they may
not fly or feed until spring, when they break up into smaller

groups and move into "summer" quarters, such as attics, crevices and cracks of buildings, and hollows of trees. With the arrival of warm weather, these bats become busy hunters and feed on insects from dusk to dawn. Once one of these bats has its "sights" on an insect, it darts and dives, dips and dodges, until the insect is caught. The bat may either eat the victim while still in flight, or more likely, roost somewhere to eat it.

The little brown bat frequently is seen in open fields, fig. 2, and may be found in abundance near lakes and other bodies of water.

Bats of this species arrive at the "summer" colonies early in April. Females are then already pregnant; they produce young between mid-May and mid-July. A female gives birth to a single young, which she carries with her for a few nights and later leaves in the roost while she goes forth to feed. The young is able to fly and fend for itself when about a month old. Toward the end of summer, individuals of this species store up great quantities of fat beneath the skin, to be used during the winter sleep. In the fall they gradually desert the "summer" colonies and by mid-November all have retreated to hibernating quarters.

Frequently many generations of the little brown bat use the same cave to hibernate in or the same building to roost in during the summer. Bats of this species possess a definite homing instinct and will return many miles to their colonies.

Distribution.—The little brown is the bat most common in Illinois. However, from some localities it is seemingly absent in summer, its place being taken by the evening bat. The Illinois specimens are of the subspecies *Myotis lucifugus lucifugus* (Le Conte). The species occurs from Labrador to southern Alaska and southward to southern California, northern Oklahoma, and southern Georgia.

MYOTIS AUSTRORIPARIUS (Rhoads)
Southeastern Bat

Description.—The southeastern bat is of medium size and has large feet. It is grayish brown on the upper parts and pale gray or white on the belly. The hairs of its dense woolly fur lack burnished tips. The nose of a live individual of this species is flesh colored, whereas the noses in other species of *Myotis* with which *austroriparius* might be confused are brown or black.

Length measurements: head and body about 2⅛ inches (53–55 mm.); tail 1⅛–1⅝ inches (28–40 mm.); over-all 3¼–3¾ inches (81–95 mm.); hind foot about ⅜ inch (10–11 mm.); ear ½–⅝ inch (13–15 mm.).

The skull is long, slender, rounded behind, has a slight sagittal crest, and attains a maximum length of 14.0–15.5 mm. (about ⅝ inch). Dental formula: I 2/3, C 1/1, Pm 3/3, M 3/3.

The southeastern bat is difficult to distinguish from related species unless specimens are available for comparison. The best distinguishing characters are woolly fur, grayish brown back, pale gray or white belly, and large feet, each 10–11 mm. in length.

Life History.—Little information about the life history of the southeastern bat is available. In Illinois, this bat is known to spend the winter months in caves and mines in the southern part of the state. Bats of this species characteristically hibernate in large, dense clusters.

Distribution.—The southeastern bat is known in Illinois only from caves and mines in Alexander and Hardin counties. Specimens have not been found in summer, and it is not known how common or widespread the species is in this state. The subspecies in Illinois is *Myotis austroriparius mumfordi* Rice. The range of the species is poorly known, but specimens have been taken in southwestern Indiana, west-central Arkansas, Louisiana, Georgia, and Florida, as well as southern Illinois, fig. 50.

MYOTIS GRISESCENS Howell
Gray Bat

Description.—The gray bat, which is larger than the little brown bat, is the only species of *Myotis* having a forearm 42 mm. or more in length. The attachment of the wing membrane extends on the leg only as far as the ankle or tarsus. The fur is velvety, and each hair is of the same gray color from base to tip.

Length measurements: head and body 2–2¼ inches (51–57 mm.); tail 1¼–1¾ inches (33–44 mm.); over-all 3¼–4 inches (84–101 mm.); hind foot ⅜–½ inch (10–12 mm.); ear about ⅝ inch (15–16 mm.).

The skull, slightly larger than that of any other *Myotis* in Illinois, has an over-all length of 15.9–16.4 mm. (about ⅝ inch); it has a pronounced sagittal crest. Dental formula: I 2/3, C 1/1, Pm 3/3, M 3/3.

Life History.—In summer, the gray bat sometimes congregates in large numbers in limestone caverns of this state. With the approach of cold weather, most bats of this species leave these caverns for places unknown. A few, however, may hibernate in the Illinois caverns. Very little is known about this bat in Illinois, for it has been taken at only a few places.

Distribution.—The gray bat is known in Illinois only from Pike and Hardin counties, but it likely occurs throughout the southern half of the state. No subspecies has been named. The range of the species extends from eastern Kentucky and Tennessee to western Missouri and northeastern Oklahoma, with an extension in the east to northwestern Florida, fig. 50.

MYOTIS KEENII (Merriam)
Keen's Bat

Description.—Keen's bat is similar to the little brown bat except that the ears are longer, fig. 51*b*. When an ear is laid forward, it extends 3 or 4 mm. beyond the tip of the nose. The tragus in each ear is narrower and longer (about 9 mm. rather than about 6 mm. long). The color of the fur on the under parts is a buffier gray than in the little brown bat.

Length measurements: head and body about 1⅞ inches (47–48 mm.); tail 1⅜–1¾ inches (36–43 mm.); over-all 3¼–3½ inches (84–90 mm.); hind foot ⅜ inch (8–9 mm.); ear from notch ⅝ inch (15–18 mm.).

The skull is essentially as in the little brown bat. Its length is 15–16 mm. (about ⅝ inch). Dental formula: I 2/3, C 1/1, Pm 3/3, M 3/3.

Life History.—In Illinois, Keen's bat hibernates in caves and mines. It is less gregarious than other species of *Myotis*; groups of a few individuals each hang apart from the larger clusters of other kinds of bats in the hibernating chambers. Although little is known of its "summer" habits, this bat apparently is not colonial, and individuals find suitable abodes in such places as mines, attics, crevices of buildings, and under eaves of houses.

Distribution.—Keen's bat, distributed the length and breadth of Illinois, may be common in some places, but the records of the species for this state are surprisingly few. The subspecies in Illinois is *Myotis keenii septentrionalis* (Trouessart). The range of the species is discontinuous, one population occurring in an area extending from Newfoundland to western North Dakota and southward to central Arkansas and western Florida and another population occurring in western British Columbia and western Washington.

MYOTIS SODALIS Miller & Allen
Indiana Bat

Description.—The Indiana bat is similar to the little brown bat in body and skull size. The fur of the back is composed of hairs that are blackish brown at bases, dull pinkish gray at tips, rather than a bronzy brown as in the little brown bat. The pinkish gray is conspicuous on the fur of the under side. The Indiana bat differs from the little brown bat in that the hind foot is smaller (the foot of the latter is 10–11 mm. in length), the hairs on the toes are much shorter and less conspicuous, and along the calcar is a keel, or flap of skin, as shown in fig. 39. The keel is most evident in live animals; frequently, in dry study skins it cannot be discerned.

Length measurements: head and body 1½–1⅞ inches (40–48 mm.); tail 1¼–1⅝ inches (30–42 mm.); over-all 2¾–3½ inches (70–90 mm.); hind foot ⅜ inch (7–9 mm.); ear from notch ⅜–⅝ inch (10–15 mm.).

The skull differs from that of the little brown bat in such small details as having a narrower braincase (less than 7.4 mm., about ¼ inch) and narrower interorbital breadth (less than 4.0 mm., about ⅛ inch). The number of teeth is the same as in other species of *Myotis*. Dental formula: I 2/3, C 1/1, Pm 3/3, M 3/3.

Life History.—The Indiana bat usually hibernates in caves. In these places individuals hang in compact clusters, often near groups of the little brown bat. Hundreds of them are neatly packed together in each cluster. Little is known about the colonies after they leave the hibernating caves. However, some Indiana bats may spend the "summer" in small colonies in other caves.

Distribution.—The Indiana bat is uncommon in Illinois, although it is abundant in southern Indiana. It has been taken in Hardin, La Salle, and Jo Daviess counties. No subspecies of the Indiana bat has been named. The range of the species extends from New England to southwestern Wisconsin and southward to central Arkansas, northern Alabama, and western North Carolina, fig. 50.

MYOTIS SUBULATUS (Say)
Small-Footed Brown Bat

Although the small-footed brown bat has not been reported in Illinois, there are many records of this bat in the northeastern United States and isolated records for Missouri, Iowa, and Kentucky. It is therefore possible that it may be discovered in Illinois when the caves and mines of this state are thoroughly investigated.

Among the distinguishing features of this species are the unusually small feet and the golden sheen of the fur.

LASIONYCTERIS NOCTIVAGANS (Le Conte)
Silver-Haired Bat

Description.—The silver-haired bat, fig. 53, is black or blackish brown, with white, appearing like silver, at the tips of many of the hairs on its back. Fur continues from the back onto the tail membrane and halfway to its tip. The ears are short and rounded and each has a blunt tragus.

Length measurements: head and body $2\frac{1}{8}$–$2\frac{1}{2}$ inches (54–64 mm.); tail about $1\frac{1}{2}$–2 inches (38–48 mm.); over-all $3\frac{5}{8}$–$4\frac{3}{8}$ inches (92–112 mm.); hind foot $\frac{3}{8}$–$\frac{1}{2}$ inch (9–12 mm.); ear from notch $\frac{5}{8}$ inch (14–16 mm.).

The skull, slightly larger than in *Myotis,* has a length of 16–17 mm. (about $\frac{5}{8}$ inch). Dental formula: I 2/3, C 1/1, Pm 2/3, M 3/3.

The silver-haired bat can be distinguished from the big brown bat by the blackish fur with "silver-tipped" hairs on its back and by its smaller size; from the red bat and the hoary bat by its darker color and incompletely furred tail membrane.

Life History.—The silver-haired bat is known in Illinois mainly as a migrant. A few individuals winter in the southern

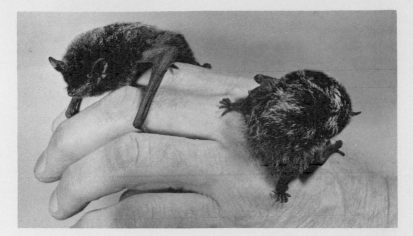

Fig. 53.—Silver-haired bat.

part of the state, but most of them migrate farther south. An inhabitant of woodlands, fig. 1, this bat lives alone (not in colonies) in the hollows of trees, beneath bark, or under leaves. It often pursues insects near the tops of trees. It may occasionally breed in Illinois, particularly in the northern part. An adult female usually bears two young, in June or July, which must develop rapidly to be prepared for the strenuous fall migration southward.

Distribution.—The silver-haired bat is moderately common throughout Illinois during the summer months. About the last week of April it is fairly common in the vicinity of Urbana. No subspecies is recognized. The species occurs from Nova Scotia and southern Quebec to southern Alaska and southward to central California, southern Arizona, southern Kansas, southern Alabama, and southern South Carolina.

PIPISTRELLUS SUBFLAVUS (Cuvier)
Eastern Pipistrel

Description.—The eastern pipistrel, fig. 54, is the smallest Illinois bat. Its fur is yellowish brown to drab brown, lighter on the under parts of the body, and the wings are dark reddish brown. Each ear has a blunt but straight tragus. This pipistrel resembles the red bat but is paler and much smaller.

Length measurements: head and body about 1¾ inches (44–47 mm.); tail 1⅛–1½ inches (29–39 mm.); over-all about 2⅞–3⅜ inches (73–86 mm.); hind foot about ⅜ inch (8–11 mm.); ear from notch ⅜–⅝ inch (9–15 mm.). Weight: about a sixth of an ounce (4.2 gm.).

The skull is small, measuring 12.3–13.0 mm. (about ½ inch) in length. Dental formula: I 2/3, C 1/1, Pm 2/2, M 3/3.

Life History.—The eastern pipistrel spends the winter in deep sleep, individuals hanging singly or in clusters of several in the dark, damp chambers of caves. Because of the moisture in the caves, droplets of water often cover the fur, and, when a beam of light is played on the bats, the fur appears to glitter like a coat of pearls, fig. 54..

Fig. 54.—Eastern pipistrel.

Toward spring many pipistrels desert the winter caves and take up abodes in attics of buildings, under porches, and in other caves. Apparently they return to the same "summer" roosts year after year. If removed some distance from such a roost, they exhibit a definite homing instinct to return to it. During the "summer" months, eastern pipistrels are entirely social and usually hang in clusters.

In early summer, the sexes tend to separate while the females bring forth their young. Usually an adult female has two off-spring.

Pipistrels are early evening flyers and frequently hunt insects before sundown. Their flight appears to be very erratic.

Distribution.—The eastern pipistrel is common in much of Illinois except in the Chicago area. It seems to be absent from the Urbana region. It is more common in the southern than the northern half of the state. Presumably, two subspecies occur in this state, *Pipistrellus subflavus obscurus* Miller in the north-western part and *P. s. subflavus* (Cuvier) in the remainder of Illinois. The range of the species extends from New England to Minnesota (except for Michigan and northern Indiana) and southward as far as eastern Mexico and central Florida.

EPTESICUS FUSCUS (Beauvois)
Big Brown Bat

Description.—The big brown bat, fig. 55, as the name implies, is both big and brown. It has a wingspread of about a foot but it weighs only half an ounce. It is dark or bronzy brown, except for blackish ears, wings, and tail membrane. Its ears are short and its tragi blunt.

Length measurements: head and body 2⅜–2⅞ inches (60–72 mm.); tail 1⅜–1⅞ inches (35–48 mm.); over-all 3¾–4¾ inches (95–120 mm.); hind foot about ½ inch (10–14 mm.); ear from notch about ¾ inch (16–20 mm.). Weight: about ½ ounce (13–16 gm.).

The skull is comparatively large and broad; it is 18.0–20.5 mm. (about ¾ inch) in length. Dental formula: I 2/3, C 1/1, Pm 1/2, M 3/3.

Life History.—The big brown bat, common in Illinois in summer, spends the daylight hours in a variety of places: in attics of dwellings; in hollows of trees; beneath boards, shutters, or awnings; in abandoned buildings and in chimneys. An adult female produces one or two young in early summer.

Being larger than most other bats, this kind includes in its diet some of the larger insects, particularly June beetles and click beetles, mayflies, caddisflies, lacewings, and parasitic wasps. It is a late feeder, becoming most active when twilight is fading into darkness.

In late summer, the big brown bat becomes exceptionally fat; the fat gives a reserve of energy for the hibernation period ahead. This bat winters in caves and buildings; it hangs less often in clusters than does the little brown bat or the Indiana bat. Apparently it is more tolerant of cold than are most other

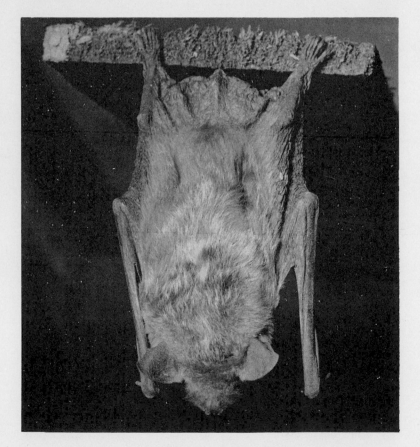

Fig. 55.—Big brown bat.

hibernating bats, and individuals may completely arouse themselves in midwinter and even fly abroad.

Distribution.—The big brown bat occurs throughout Illinois. The subspecies in this state is *Eptesicus fuscus fuscus* (Beauvois). The species is widely distributed in North America, ranging from Nova Scotia and southern Quebec in eastern Canada almost to the Yukon in western Canada and southward to Panama. The range includes all of the United States except parts of Florida and extreme southern Georgia and Alabama.

LASIURUS CINEREUS (Beauvois)

Hoary Bat

Description.—The hoary bat, fig. 56, with a wingspread of more than 13 inches, is one of the largest bats in North America. The fur is a mixture of bright buff, yellows, and deep amber, with a frosting of white laid in waves over the back. This white frosting gives the bat a decidedly hoary appearance. On the throat the fur is more yellowish and on the abdomen more whitish than on the back. The tail membrane is completely furred and much the same color as the back. The ears are short, rounded, and thickened at the rims.

Length measurements: head and body 3–3⅜ inches (77–86 mm.); tail 2⅛–2½ inches (53–64 mm.); over-all 5⅛–5⅞ inches (130–150 mm.); hind foot about ½ inch (10–14 mm.): ear from notch about ¾ inch (17–19 mm.).

The skull is large and particularly broad, as in fig. 40*e*; its length is 17–20 mm. (about ¾ inch). Dental formula: I 1/3, C 1/1, Pm 2/2, M 3/3.

Life History.—The hoary bat normally winters in states farther south than Illinois and migrates northward in spring. Some individuals spend the "summer" season in Illinois and some pass through this state while going farther north. In the "summer" months, the hoary bat spends the daylight hours in trees, among the leaves or on the trunks.

The hoary bat tends to be solitary; usually each individual hangs alone and feeds with only one or two others. Bats of this species are powerful fliers, and their slower, longer wing beats undoubtedly permit them to have a more extensive cruising range than other bats. Usually they start their evening hunting fairly late, but sometimes they fly before sundown. Although bats have relatively few predators, it is known that a long-eared owl in Illinois fed on two hoary bats.

Normally an adult female gives birth to two young in late May or early June and a few weeks later she may be so bur-

Fig. 56.—Hoary bat.

dened with the partially grown clinging young that she finds it difficult to fly.

The hoary bat is somewhat similar to the red bat and the silver-haired bat in its migration and solitary habits and in its habitat preferences.

Distribution.—The hoary bat apparently is quite rare in Illinois, although there are summer records of it from all parts of the state. The Illinois subspecies is *Lasiurus cinereus cinereus* (Beauvois). The known range of the species includes most of North America: from Nova Scotia and southern Quebec northwestward almost to the Yukon and southward to include most of Mexico. It does not include Lower California, part of southern Mexico, and the southern half of Florida.

LASIURUS BOREALIS (Müller)
Red Bat

Description.—The red bat, fig. 57, is the most colorful of the bats found in Illinois. The brick or rusty red color of its body, with a dusting of white, clearly distinguishes it from all other kinds. The wings are dark reddish brown and the shoulder patches light brown. The female is of a slightly paler color than the male. This bat is of medium size, the ears are short and rounded, and the tail membrane is fully furred on the top surface.

Length measurements: head and body 2⅛ inches (55 mm.); tail 1½–2⅛ inches (40–55 mm.); over-all 3¾–4¼ inches (95–110 mm.); hind foot about ⅜ inch (8–10 mm.); ear from notch about ½ inch (10–13 mm.).

The skull is 12.8–14.2 mm. (about ½ inch) in length. Dental formula: I 1/3, C 1/1, Pm 2/2, M 3/3.

Life History.—In Illinois the red bat is an early spring and late fall migrant and a common summer resident. In the late spring, a female red bat with one or two young attached may be found on leaves of trees, on the ground, or on house porches; the mother may be so undernourished and the young so large that she cannot become air-borne.

Bats of this species are solitary and spend the daytime of "summer" months beneath bark, under leaves, or in cracks of

Fig. 57.—A family of young red bats hanging with heads down from the branch of a tree (as viewed from directly below).

trees, or even among tall weeds and shrubs. It is thought that
the coloration of the red bat, particularly the dusting or frosting
of white, enables it to remain better hidden in less protected
places than some other bats.

The long, narrow wings of the red bat make it an excellent
flyer and an adept insect catcher. Sometimes members of this
species start foraging beneath the branches of trees even before
the sun has gone down, but normally they become active about
twilight. They feed on insects around trees and other objects,
including cribs containing corn that is heavily infested with
grain moths.

On one occasion in eastern Illinois, several red bats were at-
tracted by the watery appearance of freshly tarred roads, or by
insects caught in the tar, and were trapped in the sticky material.
At Starved Rock State Park in Illinois, a red bat and a blue
jay were captured together, the bat fastened by its teeth to the
side of the jay's head; the bird was nearly exhausted, but the bat
appeared neither exhausted nor hurt.

The red bat frequently migrates southward in groups. It is
not known whether the "summer" residents of extreme southern
Illinois migrate out of the state or whether they overwinter
there.

Distribution.—The red bat is common over all of Illinois
during the summer months. The subspecies in this state is
Lasiurus borealis borealis (Müller). The range of the red bat
includes most of the North American continent from extreme
southern Canada to Panama. Apparently not in the range is a
broad, irregular area that extends from central Mexico through
New Mexico into British Columbia and includes part of eastern
California and all of Oregon, Washington, and Idaho.

NYCTICEIUS HUMERALIS (Rafinesque)

Evening Bat

Description.—The evening bat appears to be a miniature of
the big brown bat; it is bronze or chocolate-brown, except for
blackish ears, wings, and tail membrane. The wingspread is
about 10 inches, and the length of each forearm is less than
1⅝ inches (40 mm.).

Length measurements: head and body 1¾–2½ inches (46–62
mm.) ; tail 1⅜–1⅝ inches (34–42 mm.) ; over-all 3⅛–4⅛ inches

(80–104 mm.); hind foot ⅜–½ inch (9.5–14 mm.); ear from
notch ½–⅝ inch (13–17 mm.).

The skull is short, broad, particularly between the eyes, and
unique among Illinois bats in having a single incisor, canine, and
premolar in each side of the upper jaw. Dental formula:
I 1/3, C 1/1, Pm 1/2, M 3/3.

The evening bat is frequently confused with the big brown
bat, but it can be readily told by its smaller size and its shorter
forearms (about 1⅜ inches or 35–37 mm.). It differs from the
little brown bat and other species of *Myotis* in having blunt and
curved (rather than pointed and straight) tragi, only six teeth
in each side of the upper jaw (rather than nine), and a skull
that is broader through the interorbital region.

Life History.—In summertime, the evening bat seeks refuge
in buildings, foliage, or hollows of trees. Apparently it is
colonial. The absence of winter records for Illinois indicates
that bats of this species migrate southward. The evening bat
has a feeding and flight behavior that is said by some observers
to be much like that of the little brown bat; the flight of the
evening bat is said by other observers to be more steady and
straight. Members of this species commence feeding early in
the evening, often while the sun is still shining. A female usuallv
gives birth to a pair of young each year.

Distribution.—The evening bat is abundant in southern
Illinois and at least as far north as Urbana; also, it is known
in the northeastern part of the state. The subspecies occurring
in Illinois is *Nycticeius humeralis humeralis* (Rafinesque). The
known range of the species extends from eastern Maryland to
southern Michigan and southwestward and southward to northeastern Mexico and southern Florida, fig. 50.

CORYNORHINUS RAFINESQUII (Lesson)
Southeastern Big-Eared Bat

Description.—The southeastern big-eared bat, fig. 58, is
grayish brown or smoky brown. The under parts are silvery,
not buffy as in the closely related western big-eared bat, *Corynorhinus townsendii*. The enormous ears are nearly 1½ inches
long and are joined near their bases above the forehead. Lumps
are present on each side of the muzzle in front of the eyes. The
top side of the tail membrane is not furred.

Length measurements: head and body about 2 inches (48–52 mm.) ; tail 1¾–2⅛ inches (44–54 mm.) ; over-all 3⅝–4⅛ inches (92–106 mm.) ; hind foot about ½ inch (10–13 mm.) ; ear from notch 1¼–1½ inches (33–38 mm.). Weight: less than half an ounce (about 12 gm.).

The skull is of medium size (over-all length about 16 mm., or ⅝ inch). Dental formula: I 2/3, C 1/1, Pm 2/3, M 3/3.

Fig. 58.—Southeastern big-eared bat.

The large, joined ears and lumps on the nose serve to distinguish this bat from all other kinds in Illinois. Until recently this bat was known by the scientific name *Corynorhinus macrotis* (Le Conte).

Life History.—Little is known of the southeastern big-eared bat in Illinois. It is known to live in winter and summer in caves or in structures having cave conditions. Individuals live singly or in small colonies; the females when gravid live in colonies separate from those of the males. An adult female usually produces one young in early summer; the young and mother remain together for a period in a nursery colony. Although the big-eared bat appears to be a strong flyer, it apparently does not undertake long migrations and may hibernate in the vicinity of the summer colonies. The large ears of this bat suggest that they are unusually effective receivers of sound.

Distribution.—The southeastern big-eared bat is known in Illinois only from Wabash and Alexander counties; it is evidently rare in this state. The range of this bat extends from southern Virginia and western West Virginia westward through southern Indiana and southern Illinois and southward as far as southern Louisiana and central Florida, fig. 50.

ORDER CARNIVORA
Carnivores or Flesh Eaters

The name Carnivora means flesh eaters. Members of the order are usually referred to as carnivores. Most of them are known as predators because they kill and eat other animals. Domestic dogs and cats belong to this order, as do bears, weasels, otters, skunks, raccoons, and several other wild animals.

Many of the carnivores are large animals which have vanished from densely populated areas such as Illinois and occur now only in wilder and more sparsely settled parts of the country. The black bear, the timber wolf, the red wolf, and the mountain lion are among those which once roamed this state but no longer live here. The known distribution in the United States of the least weasel and the badger, two carnivores having a limited range in Illinois, is shown in fig. 59.

Domestic dogs and cats are extremely similar in skeletal characters to their wild relatives. For this reason they are included in the identification keys prepared for this order.

Economic Status.—Wild animals of the order Carnivora are probably second only to those of the order Rodentia from the standpoint of economic importance in Illinois. Several carnivores are important as sources of fur. The annual income from the mink, for example, ranks second to that from the muskrat, one of the rodents. Between 1936 and 1949 the yearly intake from minks averaged about 35,000 pelts, and, at currently prevailing prices, these pelts would have a value of almost half a million dollars. The furs of other carnivores, such as raccoons, skunks, badgers, and red foxes, are of little value currently, but they have been valuable in the past and they may be again when fashion demands long-haired furs.

Because of their predation on rodents and rabbits, weasels, minks, badgers, and foxes are undoubtedly of great monetary value to the Illinois farmer. Skunks consume enormous quantities of insects during the summer months. Raccoons and foxes provide sport to thousands of Illinois residents and a certain aesthetic appeal to many people.

On the other hand, some carnivores—notably raccoons, weasels, minks, skunks, coyotes, and foxes—prey to some extent on domestic poultry and game birds or their eggs. Although the loss

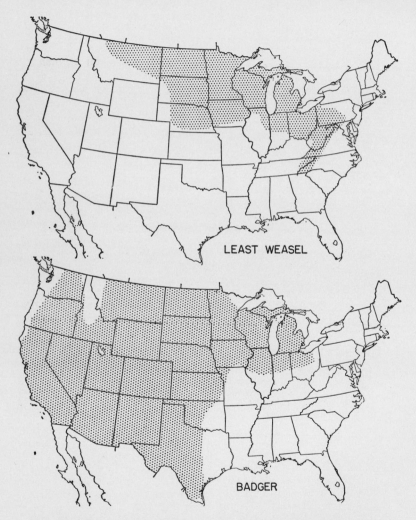

Fig. 59.—Known distribution, in the United States, of two Illinois carnivores with a limited range in Illinois.

from such predation is less than is commonly supposed, it has
made these animals unpopular in some areas and among certain
groups of Illinois residents. Destruction of game bird eggs and
young by rodents is sometimes attributed to carnivores.

KEY TO SPECIES

Whole Animals

1. Hind foot and front foot each with 5 toes, all of them with
   claws; plantigrade.   2
   Hind foot with 4 toes, front foot with 5 toes, all of them
   with claws; the "thumb" situated well above the level of
   the other toes; digitigrade.   13
2. Tail with rings of buff and black; face with black mask
   across forehead and eyes, fig. 62. . . . . . .raccoon, *Procyon lotor*
   Tail without rings; face with no dark mask.   3
3. Toes fully webbed; tail long and thickened at base. . . . .
   . . . . .river otter, *Lutra canadensis*
   Toes not fully webbed; tail not thickened at base.   4
4. Forehead with a median white stripe or square patch.   5
   Forehead with no median white marking.   7
5. Body black or black with conspicuous white markings; tail
   about ½ length of head+body.   6
   Body brownish or yellowish gray; tail about a sixth of
   length of head+body. . . . . .badger, *Taxidea taxus*
6. Upper side of body with 2 pairs of broken white stripes. . . .
   . . . . .spotted skunk, *Spilogale putorius**
   Upper side of body with 1 pair of white stripes or with a
   prominent white patch on nape and back of head, fig. 67
   . . . . .striped skunk, *Mephitis mephitis*
7. Over-all length of animal more than 1,000 mm. (39 in.);
   tail very short; last molar larger than preceding tooth. .
   . . . . black bear, *Ursus americanus*†
   Over-all length of animal less than 1,000 mm.; tail at least
   a sixth of length of head+body; last molar smaller than
   preceding tooth.   8
8. Cheek teeth 5 above, 6 below on each side of jaws.   9
   Cheek teeth 4 above, 5 below on each side of jaws.   10
9. Skull less than 90 mm. (3½ in.) long . . . .
   . . . . .pine marten, *Martes americana*†
   Skull more than 90 mm. long. . . . . .fisher, *Martes pennanti*†
10. Under parts of animal brown except occasionally for small
    light patches on throat and belly; over-all length of ani-
    mal more than 410 mm. (16 in.). . . . . . .mink, *Mustela vison*
    Under parts of animal white or yellow; over-all length of
    animal less than 410 mm.   11

*This species may occur in Illinois, but there are no official records of it.
†This species is no longer found in Illinois.

11. Tail not black tipped, less than 40 mm. (1½ in.) long; over-
     all length of animal less than 220 mm. (8⅝ in.)
     ....least weasel, *Mustela rixosa*
     Tail black tipped, more than 40 mm. long; over-all length
     of animal more than 230 mm. (9 in.)................. 12
12. Feet white; tail less than a third of over-all length of animal
     ....short-tailed weasel, *Mustela erminea**
     Feet predominantly brown; usually tail more than a third
     of over-all length of animal...........................
     ....long-tailed weasel, *Mustela frenata*
13. Claws retractile, nearly concealed in fur and strongly
     curved, fig. 61e; total number of teeth 30 or less....... 14
     Claws not retractile, but well exposed and only moderately
     curved, fig. 61f; total number of teeth 32 or more...... 16
14. Tail less than a third of over-all length of animal; upper
     jaw with 3 cheek teeth on each side.....bobcat, *Lynx rufus*
     Tail at least a third of over-all length of animal; upper
     jaw with 4 cheek teeth on each side.................... 15
15. Skull more than 175 mm. (6⅞ in.) long; over-all length of
     animal more than 800 mm. (31½ in.)..cougar, *Felis concolor*†
     Skull less than 100 mm. (about 4 in.) long; over-all length
     of animal less than 700 mm. (27½ in.)................
     ....common cat, *Felis domestica*
16. Pupil of eye of living animal elliptical as in cats; length
     of head+body less than 30 inches; tail straight and
     bushy 17
     Pupil of eye circular; tail and length of head+body varia-
     ble but, if head+body less than 30 inches long, tail usu-
     ally curled or short haired........................... 18
17. Body gray above; tip of tail black; back of ears black......
     ....gray fox, *Urocyon cinereoargenteus*
     Body reddish yellow above; tip of tail usually white; back
     of ears rusty yellow...   red fox, *Vulpes fulva*
18. Tail variable in shape, size, and color, usually not black
     tipped and without a black stripe along upper surface;
     nose and ears variable......domestic dog, *Canis familiaris*
     Tail long, bushy, with black tip and tendency toward a
     black stripe on the upper surface; nose and ears pointed.. 19
19. Legs, snout, and back of ears reddish; upper surface of
     body distinctly reddish and with gray frosting..........
     ....red wolf, *Canis niger*†
     Legs and face less reddish; upper surface of body gray and
     without reddish cast.................................. 20
20. Nosepad more than 24 mm. (1 in.) wide; over-all length of
     animal usually more than 1,400 mm. (55 in.)............
     ....timber wolf, *Canis lupus*†
     Nosepad less than 24 mm. wide; over-all length of animal
     less than 1,300 mm. (51 in.)...........coyote, *Canis latrans*

---

*This species may occur in Illinois, but there are no official records of it.
†This species is no longer found in Illinois.

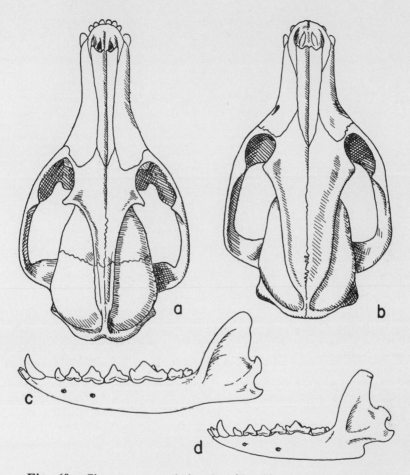

Fig. 60.—Characters used in the identification of carnivores: *a,* skull of red fox, top view; *b,* skull of gray fox, top view; *c,* lower jaw of red fox, side view; *d,* lower jaw of gray fox, side view.

Skulls

1. Teeth 30 or fewer... 2
   Teeth 34 or more... 4
2. Teeth 28; 14 in each jaw................bobcat, *Lynx rufus*
   Teeth 30; 16 in upper jaw................................. 3
3. Skull length more than 175 mm. (6⅞ in.)................
   .... cougar, *Felis concolor*†

   †This species is no longer found in Illinois.

Skull length less than 100 mm. (4 in.)..................
.....................................common cat, *Felis domestica*
4. Teeth 40 or more....................................... 5
   Teeth 38 or fewer..................................... 11
5. Teeth 40; 2 molars on each side of lower jaw............
   ....raccoon, *Procyon lotor*
   Teeth 42; 3 molars on each side of lower jaw............ 6
6. Rostrum short, the anterior end of the nasals midway between the orbit and the base of the incisors; last upper
   molar larger than preceding molar....................
   ....black bear, *Ursus americanus*†
   Rostrum long, the anterior end of the nasals much closer to
   the incisors than to the orbit, fig. 60; last upper molar
   smaller than preceding tooth......................... 7
7. Frontal bones each with a depression along mesal side of
   postorbital process, fig. 60; skull length less than 150 mm.
   (5⅞ in.) 8
   Frontal bones each convex, rounding over postorbital process; skull length usually more than 150 mm............ 9
8. Temporal ridges widely separated posteriorly and lyrate in
   shape, fig. 60*b*; inferior margin of lower mandible with
   poster or "step" or third notch, fig. 60*d* ....
   ....gray fox, *Urocyon cinereoargenteus*
   Temporal ridges close together posteriorly, fig. 60*a*, or absent; inferior margin of lower mandible without posterior "step," fig. 60*c*...............red fox, *Vulpes fulva*
9. Width of skull across zygomatic arches usually more than
   105 mm. (4⅛ in.) ; length of skull usually more than 215
   mm. (8½ in.)..
   ....timber wolf, *Canis lupus*†, or red wolf, *Canis niger*†
   Width of skull across zygomatic arches usually less than 105
   mm.; length of skull usually less than 215 mm.......... 10
10. Palatal width divided by alveolar length of upper premolar-
    molar series equaling 0.32 or less; rostrum width less than
    18 per cent of skull length‡............coyote, *Canis latrans*
    Palatal width divided by alveolar length of upper premolar-
    molar series equaling 0.33 or more; rostrum width more
    than 18 per cent of skull length‡ ....
    ....domestic dog, *Canis familiaris*
11. Upper jaw with 5 cheek teeth on each side............... 12
    Upper jaw with 4 cheek teeth on each side............... 14
12. Lower jaw with 5 cheek teeth on each side ....
    ....river otter, *Lutra canadensis*
    Lower jaw with 6 cheek teeth on each side 13
13. Length of skull less than 90 mm. (3½ in.) ....
    ....pine marten, *Martes americana*†
    Length of skull more than 90 mm..... fisher, *Martes pennanti*†

†This species is no longer found in Illinois.
‡These characteristics are evident in relatively pure-blooded animals but plainly will not distinguish between animals which are less than pure blooded.

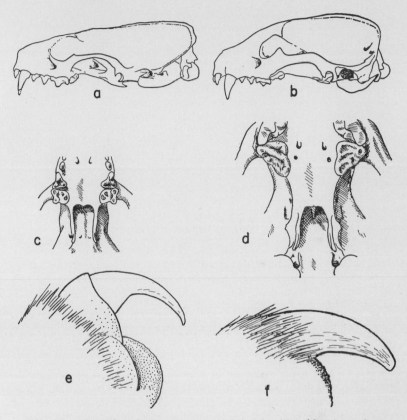

Fig. 61.—Additional characters used in the identification of carnivores: *a,* skull of mink, side view; *b,* skull of badger, side view; *c,* palatal bone of skull of spotted skunk, under side; *d,* palatal bone of skull of badger, under side; *e,* claw of domestic cat; *f,* claw of dog.

14. Hard palate extending posteriorly to about even with last molar, fig. 61*c* 15
    Hard palate extending well behind last molar, fig. 61*d* 16
15. Interorbital region of skull almost flat . . . .
    . . . . spotted skunk, *Spilogale putorius**
    Interorbital region of skull strongly convex . . . .
    . . . . striped skunk, *Mephitis mephitis*
16. Length of skull more than 100 mm. (4 in.); in lateral view,
    the anterior portion of each zygomatic arch twice as broad
    as the posterior portion, fig. 61*b* . . . . . badger, *Taxidea taxus*
    Length of skull less than 100 mm.; in lateral view, the an-

*This species may occur in Illinois, but there are no official records of it.

terior and posterior portions of the zygomatic arches approximately equal in width, fig. 61*a*.................... 17
17. Length of skull more than 55 mm. (2⅛ in.) ; auditory bullae about as long as upper premolar-molar series in each row ....mink, *Mustela vison*
    Length of skull less than 55 mm.; auditory bullae longer than upper premolar-molar series...................... 18
18. Length of skull less than 33 mm. (1¼ in.)................
    ....least weasel, *Mustela rixosa*
    Length of skull more than 33 mm........long-tailed weasel, *Mustela frenata,* and short-tailed weasel, *M. erminea**

URSUS AMERICANUS Pallas
Black Bear

Description.—The black bear, the largest carnivorous animal in Illinois within historic times, is recognizable by its large size, short tail, short black or dark brown fur, and brown muzzle. On each foot it has five toes with claws.

The skull is nearly a foot long; the back teeth are broad and flat (for crushing, not cutting). Dental formula: I 3/3, C 1/1, Pm 4/4, M 2/3.

This bear is truly omnivorous. In winter, it sometimes sleeps for periods of several days, but does not truly hibernate.

Distribution.—The black bear existed in the wooded and swamp areas of Illinois during the last century. It disappeared first from the northern and central portions of the state; some individuals remained in the southern part until the latter half of the 1800's. The subspecies which occurred in Illinois was *Ursus americanus americanus* Pallas. The present range of the species includes most of Alaska and Canada, with southern extensions along the Sierra Nevada to central California, along the Rocky Mountains to the Mexican Plateau, in the great Lakes region to central Minnesota, Wisconsin, and Michigan, and along the Appalachians to northern Georgia; the range includes also Florida and the Gulf Coastal Plain.

PROCYON LOTOR (Linnaeus)
Raccoon   Coon

Description.—The raccoon, fig. 62, is a moderately large, stocky animal weighing usually 10 to 26 pounds when adult.

*This species may occur in Illinois, but there are no official records of it.

Fig. 62.—Raccoon.

It is readily distinguished by a mask of black over its eyes and by a densely furred and ringed tail of alternate black and light bands. Its fur is long and coarse; its color usually is grizzled gray-brown, but varies from yellowish gray to nearly black. Its muzzle is fairly sharp pointed, and its ears are prominent, rounded, and furred. The feet are broad and plantigrade; that is, the animal walks on nearly the entire under surface of the foot. The five toes on each foot are provided with prominent claws.

Length measurements: head and body about 20–28 inches (500–720 mm.); tail 8–11 inches (200–280 mm.); over-all 28–39 inches (700–1,000 mm.); hind foot about 4–5 inches (95–120 mm.); ear 1¾–2¼ inches (45–58 mm.).

The skull (length 105–125 mm., or about 4½ inches; width 65–80 mm., or about 3 inches) is nearly as large as that of a fox terrier dog, but has a short rostrum. The top of the skull appears to be arched when viewed from the side; a bony palate extends well behind the last molars. The teeth have short, rounded cusps adapted more for crushing than for cutting food. Dental formula: I 3/3, C 1/1, Pm 4/4, M 2/2.

Life History.—The raccoon, or coon as it is popularly called, is so common in wooded areas of Illinois that it is known to nearly everyone. It is most abundant in wooded river bottoms, fig. 3, and less abundant in wooded uplands. Watercourses without tree cover usually provide less desirable habitat, as do marshes and strip-mine ponds. Raccoons have always been moderately common in Illinois and at one time were apparently so numerous along the Kaskaskia River that the Indians are said to have used their word for the animal as their name for the river.

The raccoon usually selects for its home or den a dry cavity in a tree, fig. 1. It seems to have no preference for any particular species of tree so long as the cavity in the tree is large enough and is dry. The raccoon may also use for its den a dry crevice in a cliff, the deserted burrow of a fox or woodchuck, or even the lodge of a muskrat. It may also substitute quite satisfactorily a man-made den (a box with an opening near the top) nailed to the trunk of a tree. During the daytime, the coon sleeps in its den, and during the most severe parts of winter may remain there in a dormant state for several days. However, it does not truly hibernate as does the ground squirrel or the woodchuck, for it may awaken in a few moments; these other animals may require hours to arouse themselves fully.

At night the raccoon wanders in search of food in woods, fields, and particularly along streams or other bodies of water. It feeds on a great variety of plants and animals and is an opportunist in that it makes use of any suitable food that is at hand. In the fall, its diet may consist of three-fourths vegetable matter and one-fourth animal matter. In the spring and summer, the greater part of the diet is likely to be animal matter. Important plants in the diet are persimmons, pecans, acorns, grapes, pokeweed berries, other fruit, and corn, particularly corn in the milk stage. Important animals in the diet are insects and crayfish, and to a lesser extent snails, earthworms, birds, fish, snakes, and small mammals. When crippled waterfowl become readily available, as during and following the hunting season, the coon feeds heavily on these incapacitated birds. The coon hunts skillfully in water and is adept at digging and climbing. In the use of its hands and in its curiosity, it has a humanlike quality.

Breeding occurs in late January to early March, and the young, normally three or four to a litter, are born usually in March or April. At 30 days they occasionally leave the nest, and at about 60 days they forage with their mother.

The coon is trapped or hunted in every county of Illinois. As a furbearer, it is second or third (preceded by the muskrat and usually the opossum) in the Illinois fur catch each year. More than 10,000 trappers and hunters harvested coons in this state in 1939 and in 1940, when at least 32,000 coons are estimated to have been taken each year. Coon populations have increased since that time, and the annual catch has undoubtedly increased also. As a game mammal, the coon provides exciting sport. It is hunted during autumn nights with the help of dogs and lights. The clever running of the coon taxes the trailing ability of dogs to the utmost and furnishes exciting sport to both men and dogs.

Signs.—Coon tracks, fig. 20, are most abundant along the margins of streams, ponds, and lakes. Prints of individual feet commonly vary from 3 to 4 inches in length. Tracks of small coons look somewhat like those of large muskrats, but the latter usually are accompanied by tail marks; the birdlike prints of the four-toed front feet also distinguish the muskrat tracks. The usual gait of the coon is a walk.

Scats or droppings of coons may be found along margins of streams and lakes, usually on logs. Often great accumulations of them may be found, because coons commonly follow regular routes and leave their droppings in a few chosen spots. Large quantities of crayfish parts, corn, cherry pits, or berry seeds are characteristic of them.

Distribution.—Raccoons are present in every county of Illinois and are particularly common in those parts of the state with extensive wooded areas. Two subspecies are known to occur in Illinois, *Procyon lotor hirtus* Nelson & Goldman in the northwestern four-fifths of the state and *P. l. lotor* (Linnaeus) in the southeastern fifth. The range of the species includes extreme southern Canada; all of the United States except a western area that contains parts of the southwestern deserts, western parts of Wyoming and Montana, and most of Idaho; also the range includes Mexico and Central America.

MARTES AMERICANA (Turton)

Pine Marten   Marten

Description.—The pine marten is similar in shape to the better-known mink but is slightly larger. Its fur is yellowish brown on the body and shades to dark brown or black on the

tail and legs. On the throat and chest is a pale buff patch. The
tail (about 9 inches long) is usually half as long as the head
and body (about 18 inches).

The skull, which has a rounded, broad braincase, is approximately 85 mm. (about 3⅜ inches) long in the male and 75 mm.
(about 3 inches) in the female. Dental formula: I 3/3, C 1/1,
Pm 4/4, M 1/2.

Distribution.—The marten was recorded from Cook County
by Robert Kennicott (1855:578), and a marten skeleton taken
"in northern Illinois" is preserved in the Museum of the Chicago Academy of Sciences. Probably the marten once occurred
in several of the northern counties of our state where there
were stands of white pine. There are no records of its present-
day occurrence in Illinois, but there are several for Wisconsin.
The marten is an arboreal animal, principally of the coniferous forests. The subspecies in Illinois was *Martes americana
americana* (Turton). The species is found from Labrador,
Newfoundland, Nova Scotia, northern New York to Alaska,
and southward in western Canada to the northwestern United
States, with extensions in the Sierra Nevada to south-central
California and in the Rocky Mountains to the northern part of
New Mexico.

MARTES PENNANTI (Erxleben)
Fisher

Description.—The fisher is similar in appearance to the
marten but is larger (head and body about 23 inches long) and
darker, being mostly of a dark brown, almost black color with
a slight frosting of white. It has no buff patch on throat and
chest, although some small white patches may be present. The
tail (about 14 inches long) is somewhat longer than half the
length of the head and body.

The skull is like that of the marten but larger (about 120
mm., or about 4¾ inches, long). Dental formula: I 3/3, C 1/1,
Pm 4/4, M 1/2.

Distribution.—The fisher formerly occurred in northernmost
Illinois. According to Robert Kennicott (1855:578), it "used
frequently to be seen in the heavy timber along Lake Michigan"
in Cook County and (1859:241) "It has been found, within a
few years, in Northern Illinois, and appears to be an inhabitant

of the woods, alone." The fisher is chiefly arboreal and inhabits timbered swamps or woods near water. It feeds on various mammals, including mice and porcupines, and on birds. The subspecies in Illinois was *Martes pennanti pennanti* (Erxleben). The species now has a range from Labrador, Nova Scotia, and northern New York to northern British Columbia, with southern extensions to Yellowstone National Park and to central California.

MUSTELA RIXOSA (Bangs)
Least Weasel

Description.—The least weasel, fig. 63, as the name implies, is the smallest of our weasel-like mammals and also the smallest known American carnivore. It has a slender body (only about 1¼ inches in diameter) and a stubby tail about a fourth as long as the body. Its body is as long as that of an adult Norway rat, but is much slimmer. The least weasel has two coats or two "colors" each year, an entirely white coat and a summer coat that is dark brown above and whitish beneath. The tail is of uniform color, whether white or brown, and does not have a conspicuous black tip, although the tip may have a few darker hairs. The upper sides of the feet are white in summer as well as in winter.

Length measurements: adult male, head and body 6–7½ inches (153–191 mm.), tail 1⅛–1½ inches (28.0–37.5 mm.), over-all 7⅜–9 inches (188–230 mm.), hind foot about ⅞ inch

Fig. 63.—Least weasel.

(23 mm.) ; adult female, head and body 5–7¼ inches (127–184 mm.), tail 1–1¼ inches (25–33 mm.), over-all 6¾–8½ inches (170–217 mm.), hind foot about ¾ inch (19–21 mm.). Weight (adult) : about 2 ounces.

The skull is small (30.5–33.0 mm., or about 1¼ inches, long) and delicate. Dental formula: I 3/3, C 1/1, Pm 3/3, M 1/2.

Life History.—The least weasel commonly lives in a den about 6 inches below ground. It may appropriate the nest cavity and nest of a meadow vole. Very little is known of its breeding habits, but they are probably similar to those of the long-tailed weasel.

The least weasel can go many places that mice can go and consequently preys on them to a considerable extent. A few least weasels may be exceedingly valuable in controlling meadow voles and white-footed mice in fields. In Ohio, one storehouse of a least weasel contained nearly a hundred discarded skins of meadow voles. Probably during the summer it preys extensively on the readily available insects.

Signs.—Tracks of the least weasel are like those of other weasels but smaller and without tail marks. The toe marks rarely show. The usual gait of this weasel is a bound. The long prints of the hind feet, which lead in the bound, are between ¾ and 1 inch long, whereas those of the long-tailed weasel are between 1½ and 2 inches long.

Droppings of this weasel are packed with mouse fur and great numbers of insect parts.

Distribution.—The least weasel, uncommon in Illinois, is known to occur only in the northern three or four tiers of counties except in the eastern part of the state, where it extends south into Champaign County. The subspecies in Illinois is *Mustela rixosa allegheniensis* (Rhoads). The range of the species includes an area that extends from the St. Lawrence River northwestward to northern Alaska, with extensions southward to northern Montana, southern Nebraska, central Illinois, and along the Appalachians to North Carolina, fig. 59.

MUSTELA FRENATA Lichtenstein
Long-Tailed Weasel

Description.—The long-tailed weasel, fig. 64, is a slender animal; its length is accentuated by the long, slim tail, which is

Fig. 64.—Long-tailed weasel.

slightly bushy. In summer, its fur is a soft brown color except
for that on the abdomen and chest, which is whitish, suffused
with yellow, and the tip of its tail, which is black. In winter
in the northern part of its range, the fur of the long-tailed
weasel turns "white" except for the conspicuous black tip of the
tail; the fur of the under parts is usually washed with a yellowish color. In the southern part of the range of this animal,
both brown individuals and white individuals may be found;
those which do not become whitish have a winter coat that is
thicker than the summer coat and of a different shade of brown.
Much additional information is needed regarding the occurrence of long-tailed weasels in Illinois and the percentage of
individuals, if any, that in a particular area turn whitish.

In size, there is a marked difference between the male and
the female of the long-tailed weasel; the male is generally
considerably larger than the female. The sex of a weasel specimen should be determined before a positive identification is
attempted.

Length measurements: adult male, head and body 9¼–10¼
inches (235–260 mm.), tail 4¼–5½ inches (110–140 mm.),
over-all 13¾–15¾ inches (350–400 mm.), hind foot 1½–2
inches (40–50 mm.); adult female, head and body 8¾–9¼
inches (223–235 mm.), tail 3–4¼ inches (76–108 mm.), over-all
11¾–13¼ inches (298–338 mm.), hind foot 1¼–1½ inches
(31–38 mm.). Weight: adult male 6–10 ounces; adult female
3–7 ounces.

The skull of the long-tailed weasel is much like that of the
mink but smaller, less angular, and less strongly ridged. The

auditory bullae are inflated and long anteroposteriorly. The over-all length of the skull in the male is 43–50 mm. (about 1⅞ inches); in the female, 30–42 mm. (about 1½ inches). Dental formula: I 3/3, C 1/1, Pm 3/3, M 1/2.

Life History.—The long-tailed weasel occurs over much of Illinois in brushland, along shrubby fencerows, in haystacks, in brush piles, and sometimes under farm buildings. It preys principally on rodents and is most apt to be found where small mammals are abundant. In Illinois, as in other states, it feeds heavily upon voles, white-footed mice, ground squirrels, rabbits, rats, birds, insects, and, in fact, upon almost any animal it can overpower.

A hungry long-tailed weasel, once on the trail of a rabbit, is an intent, relentless pursuer. A healthy rabbit, however, will give the weasel a good run and may cleverly confuse its own trail by hopping back and forth over a small area. Again, the rabbit may freeze motionless and be nearly lost to sight in a tangle of brush so that the weasel may pass it by momentarily. The rabbit may make a confusing maze of tracks before dashing off, and the shorter-legged weasel must hustle to keep up with its prey. But once the gap is sufficiently narrowed between predator and prey, the weasel makes a quick dash and bites the rabbit at the base of the skull. Over and over the two may tumble, the weasel holding on tenaciously until the rabbit ceases to struggle.

The long-tailed weasel usually makes its home in the burrow of a ground squirrel or larger mammal, after slightly remodeling the entrance and nest. A few droppings near the entrance and bluebottle flies humming about announce that the burrow is no longer occupied by a seed- or herb-eater but by a carnivore that has a store of meat inside. Dry grass, leaves, and mouse fur are common nesting materials of this weasel.

An adult female gives birth annually to a single litter of young, usually four or five in number, in the spring. Throughout the first month the young remain in the den, and food is brought to them. Soon they are providing their own food. When 3 to 4 months old they are nearly full grown. An adult weasel can consume each day an amount of flesh equivalent to about a third of its weight.

Signs.—Tracks of the long-tailed weasel, like those of the mink and the least weasel, are seen usually in the bounding pattern, the long prints of the hind feet leading and the short

prints of the front feet trailing. Each footprint is oval and,
unlike the mink footprint, it rarely shows toe marks. Unlike
the tracks of the least weasel, its tracks may be accompanied
at times by tail marks. Each footprint is larger than that of
the least weasel.

Droppings of the long-tailed weasel may be found along
fences or near its den. They usually contain insect parts, fur
of mice, and bird feathers.

Distribution.—The long-tailed weasel occurs the length and
breadth of Illinois. In this state is the subspecies *Mustela frenata noveboracensis* (Emmons). The range of the species extends from southern Quebec to central British Columbia and
into Central America. Records are lacking for Lower California and some desert regions of Arizona and California.

MUSTELA ERMINEA Linnaeus
Short-Tailed Weasel

To date there are no Illinois records of the short-tailed
weasel, but since this species is known to occur at Beaver Dam,
Wisconsin, only 70 miles north of the Illinois-Wisconsin line,
there is some possibility that it may eventually be found in the
counties of extreme northern Illinois.

The short-tailed weasel is similar to the long-tailed weasel
in size and general appearance; it differs in that summer specimens have white feet. White specimens of the two species are
difficult to separate but usually differ in tail length (see key to
the Carnivora).

MUSTELA VISON Schreber
Mink

Description.—The mink, fig. 65, is weasel-like in build and
is sometimes referred to as a water-weasel. Its color is dark
brown, except for black at the tip of the tail, white on the chin,
and occasionally white spots on the throat and chest. Its ears
are short and hardly discernible in the fur. The thickly haired
tail is about two-fifths as long as the head and body. The male
is larger than the female.

Length measurements: adult male, head and body $13\frac{1}{2}$–$15\frac{3}{4}$
inches (345–400 mm.), tail $7\frac{1}{4}$–$8\frac{1}{4}$ inches (185–210 mm.),

over-all 21–24 inches (530–610 mm.), hind foot 2⅛–2¾ inches (55–70 mm.) ; adult female, head and body 11½–13⅜ inches (290–340 mm.), tail 5¼–7½ inches (135–190 mm.), over-all 16¾–21 inches (425–530 mm.), hind foot 2–2½ inches (53–62 mm.).

The skull, fig. 61a, of the mink is similar to but larger than that of the largest weasel and has a more pronounced median dorsal ridge and more flattened auditory bullae. It differs from that of the skunk in that the shelf of the bony palate extends well back of the last molar, and the last upper molar is dumbbell shaped. The skull of the male is 60–70 mm. (about 2⅝ inches) long, of the female 55–63 mm. (2⅛–2½ inches). Dental formula: I 3/3, C 1/1, Pm 3/3, M 1/2.

Life History.—The mink is at home on land or in water. It may dig a burrow for itself, or, more likely, it may take over a muskrat den in a lake or a muskrat burrow in the side of a

Fig. 65.—Mink.

stream bank for a home without remodeling it. The mink seeks food principally along the shores of lakes or the banks of streams and ditches. It feeds most often on young or sickly muskrats. It feeds also on aquatic insects, crayfish, frogs, snails, some water birds, and, since it is an excellent swimmer, it preys at times on fish, particularly the smaller kinds. Occasionally, and as circumstances dictate, it forsakes the stream or lake edge for a meadow, where it preys upon meadow mice, white-footed mice, and rabbits.

The mink hunts mainly at night. It is a bold hunter and seems to spend less time stalking its prey than do weasels. Each animal probably patrols several miles of lake shore or ditch or river bank. Like its relatives, the weasels, it kills wantonly whenever it comes upon a brood of chicks, a family of mice, or any other concentration of prey species. Its enemies are individuals of its own species, foxes, horned owls, dogs, and trappers. The mink is hunted and trapped as one of the most valuable furbearers of Illinois.

An adult female has annually but one litter of young, usually numbering 5 or 6, in April or May. At birth the young are of pea-pod size and helpless; in 2 months they may go on short forays with the mother. The brood breaks up in late summer.

The mink, like the skunk and the weasel, has scent or musk glands. Although the amount of musk discharged is small, its odor is potent and perhaps more offensive to man than that of the skunk or weasel. The scent is emitted by the mink most frequently during the breeding season, when individuals are trying to attract those of the opposite sex.

Signs.—Although the mink has five toes on each foot, only four show in each print, fig. 25. Front and hind prints are practically alike and are about 1⅜ inches across. In the usual gait of the animal, which is bounding, prints of the hind feet are almost side by side and in front, while those of the front feet are close behind these and one slightly trails the other. Six to 20 inches may separate each complete set. In fluffy snow, the tail mark may show on each set but not between sets.

Droppings of the mink may be found along the margins of waterways and may be recognized by their contents and by their shape, each having one end more or less spirally twisted. They generally are packed with hair and crayfish parts and occasionally with fish and frog remains.

In summer, a tuft of fur or feathers and bluebottle flies near what appears to be a woodchuck or a muskrat burrow are signs that a mink may be using the burrow.

Distribution.—The mink is fairly common in all parts of Illinois. The subspecies *Mustela vison letifera* Hollister is found north and west of the Illinois River and *M. v. mink* Peale & Beauvois in the remainder of the state. The range of the species includes most of Canada, Alaska, and the United States. It does not include the extreme north or the arid regions of southwestern United States.

LUTRA CANADENSIS (Schreber)
River Otter

Description.—The river otter, fig. 66, is an elongate, short-legged, thick-tailed aquatic cousin of the mink and the weasels. Its thick, short fur, small ears, round tail, which is thick at the base and tapered to a tip, and large webbed feet are adaptations for agility and speed in the water. The fur is mostly a rich dark brown, slightly lighter in tone on the under parts, grayish brown on the throat, and grayish white on the chin and lips.

Length measurements: head and body about 28 inches (700 mm.); tail 16 inches (400 mm.); over-all about 43 inches (1,100 mm.); hind foot 5⅛ inches (130 mm.). Width of tail

Fig. 66.—River otter.

at base: about 3½ inches (90 mm.). Weight: approximately 20 pounds.

The skull is large, especially the braincase, and is about 105 mm. (4 inches) long and 70 mm. (2¾ inches) wide. It is not heavily ridged. The auditory bullae are flattened, and the teeth are massive. Dental formula: I 3/3, C 1/1, Pm 4/3, M 1/2.

Life History.—The river otter lives along streams and lakes. Only infrequently does it wander far from them on excursions from one body of water to another. Its large den, usually with an entrance diameter of about 10 inches, is never more than a few hundred yards from water. Usually it is in a stream bank or on a lake shore, either above or below water, and protected by roots of large trees or by overhanging banks. For its den, the otter may use a natural cavity or a burrow, such as that of the woodchuck. Breeding occurs in winter, and about 11 months later a litter, usually of three young, is born. The male usually stays in the general vicinity of the den, but the female does not let him join the group until the young are old enough to travel. During their third or fourth month, the cubs are forced to learn to swim. A family group may hunt and fish over a waterway of 10 or more miles during the season. The river otter feeds on crayfish, frogs, turtles, earthworms, aquatic insects, and fish.

Signs.—Most conspicuous of all otter signs are the slides—the places where the animals gather to enjoy a slide down clay or snow banks into water. These slides are a foot or more wide and may be a dozen to many feet long. Otters remove every stick and stone from the slides, and, as the water dripping from their pelts makes the clay or snow very slippery, the descent is fast.

The surest place to see otter tracks, fig. 23, is on the shores and islands of large rivers. When much snow is on the ground, the otter sometimes propels itself by pushing itself along, making a deep groove in the snow; even when it walks in shallow snow its short legs allow its body to drag, leaving a continuous trail. When traveling at considerable speed through deep snow, the otter bounds in such a way that it leaves a well-marked, full-length impression of its long body on the trail.

Distribution.—The river otter was once fairly common along the large streams of Illinois, but apparently by the early 1800's it was scarce in most parts of the state. Since 1900, it has been seen or taken in 25 counties; records indicate a sporadic occurrence in all but the northeastern part of the state. The river otter in Illinois belongs to the subspecies *Lutra canadensis canadensis* (Schreber). The range of the species includes most of Canada, Alaska, and the United States.

MEPHITIS MEPHITIS (Schreber)
Striped Skunk

Description.—The striped skunk, fig. 67, is a short-legged, black and white animal, about as large as a house cat, but with a bushy tail and a pointed head. This skunk is predominantly black in color, but it always has at least a white stripe on the

Fig. 67.—Striped skunk.

forehead and a white patch on the top of the head. If an individual has no more white than this, it is classified by fur dealers and trappers as a "black" or "star." From the white patch, two white stripes may fork and extend a short distance down the back ("short stripe" or "narrow stripe"), or the stripes may extend all the way to the tail and may be joined together ("broad stripe"). The basal half of each hair on the tail is white, the other half black, except that the group of hairs at the tip of the tail may be entirely white. The claws on the front feet are long.

Length measurements: head and body 15–16½ inches (385–420 mm.); tail 7–9 inches (175–230 mm.), over-all 22–26 inches (560–650 mm.), hind foot 2¼–2¾ inches (58–70 mm.).

The over-all length of the skull is 62–80 mm. (2⅜–3⅛ inches) and the width 40–46 mm. (1½–1¾ inches). It differs from other carnivore skulls in having on each side only one upper molar, which is more or less square, and two lower molars. The auditory bullae are not greatly enlarged, and the shelf of the bony palate extends no farther than the last upper molars, approximately as in fig. 61c. Dental formula: I 3/3, C 1/1, Pm 3/3, M 1/2.

One of the characteristic things about the skunk is the odor of its musk. A yellowish material, composed chiefly of mercaptan, musk is expelled from glands, one on each side of the anus of the skunk. The musk is used by the skunk in defending itself.

Life History.—The striped skunk usually lives in a burrow or underground den—the modified home of a woodchuck or ground squirrel. It may also live beneath the floor of a barn or shed. This skunk is at home in a multitude of environments: at the edge of woods, in brushy country, along fencerows, near grassy meadows, or around outbuildings; but it seldom strays far from a drainage ditch, creek, or other source of water.

This skunk normally does its feeding at night, but occasionally it may be abroad in the daytime. It is sluggish in its movements, and usually it shows little fear of potential enemies, including man. Occasionally it may be seen ambling along in brush or woods in the daytime, fig. 2. It prefers to depend on its defensive scent mechanism rather than on rapid flight. When the skunk encounters an intruder, it first gives a warning by elevating its tail straight as an exclamation mark, pluming out its tail hair, and stamping its front feet. If the intruder comes

too close, the scent glands, with nozzles properly focused, go
into action.

The striped skunk eats whatever suitable food is most readily
available. In winter, it feeds on hibernating insects, mice, and
fruits. In the warmer seasons, it feeds heavily on berries and
other fruits, eggs, insects, mice, and some of the smaller birds
and snakes. At any season it may become a carrion eater and
probably includes in its diet many animals killed on highways.
This skunk may den up and sleep during the coldest days of
winter, but it never truly hibernates, as do some of the bats, the
woodchuck, or the ground squirrels.

Young of the skunk are born in late spring and number 4 to
10 per litter. Within a single litter, they may show much variation in the amount of white on the back; some may be "blacks"
and others "broad stripes." Although the young are blind and
helpless at birth, they grow rapidly and are hunting by July.

Signs.—The prints of the hind foot of the skunk are each
about 2½ inches long and rarely show claw marks; the prints
of the front feet are each about 2 inches long, broad, and frequently show claw marks. The track, fig. 35, made by a skunk
traveling at a walk consists of a parallel arrangement of closely
spaced footprints on each side; that by a running animal consists
of sets of footprints arranged diagonally across the direction of
travel, and in each set prints of hind feet are on the outside.

Droppings are about a half inch in diameter and frequently
contain insect parts, berry seeds, and fur. Diggings in leaf mold
or soft ground often indicate where a striped skunk has dug out
grubs or other insects or uncovered a mouse.

Distribution.—The striped skunk is moderately common in
all counties of Illinois. Two subspecies are present, *Mephitis
mephitis nigra* (Peale & Beauvois) in the southern third and extreme eastern parts of the state and *M. m. avia* Bangs in the
remainder of Illinois. The range of the species includes approximately the southern half of Canada, the United States
(except the southern tip of Florida), and northern Mexico.

SPILOGALE PUTORIUS (Linnaeus)

Spotted Skunk   Civet Cat

The spotted skunk is known from adjacent Iowa and Missouri, and there is an unreliable 1910 sight record for this ani-

mal from Golconda in southeastern Illinois. Because this skunk is in Iowa and Missouri, it may eventually be found in Illinois, perhaps in counties bordering the Mississippi River.

Diagnostic features of the spotted skunk are outlined in the key to the Carnivora.

TAXIDEA TAXUS (Schreber)
Badger

Description.—The badger is a short-legged, squat animal weighing 20 to 35 pounds, fig. 68. It is mostly yellowish gray in color and has a white stripe extending from the nose over the forehead and disappearing on the back, white cheeks, with a black spot on each, yellowish under parts and tail, and black feet. Its ears are small, its tail is short and bushy, and its large forefeet are provided with long claws. Its sturdy body and powerful forelegs make the badger an effective digger.

Length measurements: head and body about 24½ inches (620–640 mm.), tail, from base to tip of bone, 4–6 inches (100–150 mm.), over-all 28–31 inches (720–790 mm.), hind foot 3½–5 inches (90–130 mm.), ear from notch about 2 inches (50 mm.).

Fig. 68.—Badger.

The skull of the badger is broad in proportion to its length (about 120 mm. or 4¾ inches long, 85 mm. or 3⅜ inches wide), the auditory bullae are very large, and the palate extends well beyond the end of the upper molars, fig. 61*b, d*. Each lower cutting tooth (carnassial or first molar) has a prominent internal cusp. Dental formula: I 3/3, C 1/1, Pm 3/3, M 1/2.

Life History.—The badger lives in open country and seems to prefer sand prairie, as it occurs less frequently on prairies with heavier soils. It not only readily digs out the burrows of ground squirrels and woodchucks but makes its own burrows. In one case, a badger decided to "surface" from its burrow beneath a well-packed, frequently used macadamized road near Savanna; it made an opening through the macadam as if this material were no harder than soil.

In spring the badger uses its home burrow as a nest chamber in which to bring forth and care for its young, usually numbering three. In winter the animal may sleep for several days during cold periods, although it never truly hibernates.

The badger is active at night and occasionally during daylight hours. It preys extensively on ground squirrels, woodchucks, and meadow mice. Probably three-fourths of its food in Illinois consists of mammals, most of which are pests to man. It also eats insects, frogs, a few birds' eggs and nestlings, and rabbits.

Signs.—Badger tracks, fig. 36, are distinguished by long claw marks of the short, broad front feet and by the narrower prints of the hind feet. The prints of the front feet are each about 2¾ inches long and 2 inches wide, and of the hind feet about 3½ inches long and 1¼ inches wide. The usual gait of the badger is a shuffle, as shown by the parallel arrangement of the closely spaced footprints. The width between right and left prints is about 8 inches. Prints of the front feet are markedly pigeon-toed.

The burrow is 10 or more inches in diameter and usually enters the ground at a steep angle. Even shallow beginnings of badger burrows may be distinguished from the diggings of dogs; dogs tend to scratch in only one direction, while the badger rotates its body and digs at the side walls. A shallow digging by a dog tends to be vertically oval in cross section and to have relatively few claw marks on the sides and many on the bottom. The badger digging tends to be horizontally oval in outline and to have a high proportion of claw marks on the side walls.

Distribution.—According to a report by Robert Kennicott (1859:250), the badger was present in northern Illinois at the time he wrote. It had once been "numerous, at least as far south as the middle of the State," and had been seen 30 years before "near the Kaskaskia River." Subsequent reports indicate that by the late 1800's the badger had nearly disappeared from Illinois, but in recent years it has re-established itself in the northern half of the state. It is now common locally in northwestern Illinois and is known as far south as Fulton and Douglas counties. The subspecies in this state is *Taxidea taxus taxus* (Schreber). The range of the badger includes most of the western half of the United States, fig. 59, with eastern extensions into Michigan and Ohio, northern extensions into southern Canada, and southern extensions into northern Mexico.

VULPES FULVA (Desmarest)
Red Fox

Description.—The red fox is about the size of a small dog and has a pointed muzzle, prominent ears, and a long, bushy tail. It is mostly yellowish red in color (the color is more intense down the middle of the back and on the tail than elsewhere); it is whitish on the belly, throat, cheeks, and inner sides of the ears. Its feet are black, and its tail is tipped with white.

In Illinois this fox rarely shows marked color variations or color phases. A "silver" fox is an individual that has black fur, instead of red, with white hairs sprinkled among the black. A "cross" fox is an individual that has part of the fur black and part red, with the black along the back and over the shoulders forming a cross as seen from above. It resembles a gray fox superficially. A "black" is entirely melanistic.

Length measurements: head and body 28–34 inches (715–870 mm.), tail 12–15 inches (300–380 mm.), over-all 40 inches (1,015–1,025 mm.), hind foot 6¼–6⅝ inches (160–170 mm.), ear from notch 3½ inches (85–90 mm.). Weight: 8–14 pounds.

The skull is 135–160 mm. (5¼–6¼ inches) long; the two temporal ridges on top of the braincase, fig. 60*a,* may closely approach one another or may converge to form a weak sagittal crest. The lower jaw has two notches at the posterior end, fig. 60*c,* but lacks the "step" of the gray fox, fig. 60*d.* Dental formula: I 3/3, C 1/1, Pm 4/4, M 2/3.

Life History.—The red fox is so versatile and so adaptable in its food habits that it has not only been able to survive but to increase successfully in the face of constant fur and bounty hunting and a changing environment brought about by man.

This fox occupies various kinds of habitat, but it seems most at home in strongly rolling country where land use has resulted in an irregular arrangement of fields, meadows, and semiopen

Fig. 69.—Young of red fox near entrance to den.

woodlands, fig. 2. The range of an established, unharassed individual, pair, or family is not extensive, probably having a radius of about a mile.

In late December and early January, a careful observer will begin to see the double trails made by paired foxes running together, indicating that the mating season is under way. As the winter wears on, the foxes take an increasing interest in dens and by late February each pair probably will have cleaned out several in preparation for the birth of pups. Usually woodchuck and badger dens are appropriated for this purpose. The male fox may begin to leave food at the den even before the pups are born. The young generally arrive in March; the litter usually consists of four to nine.

The pups first appear outside of the den when they are about 3 weeks old. If the family is undisturbed, the pups will remain for about a month at the den in which they were born, fig. 69. The family may then move from den to den, frequently occupying as many as seven separate dens before finally deserting these underground homes. Although cubs and adults leave the last den in June, they remain together as a family group through the summer months.

The red fox is celebrated for its cleverness in obtaining food, but its prowess in this respect is often overrated. It tends to take the foods that, within the limitations of its preferences and physical abilities, are easiest to obtain.

Where many kinds of foods are available in quantity, the proportions in the annual diet are about as follows: mammals (largely rabbits and mice) 45 per cent; birds 15 per cent; insects 20 per cent; and vegetable matter (largely fleshy fruits) 20 per cent.

The occurrence of mammals in the summer and autumn diet is comparatively infrequent due to increased availability of fleshy fruits and insects in these seasons. Fruit may make up as much as 35 per cent of the diet during some warm weather months. Many kinds of insects are included in the diet, but May beetles, crickets, and grasshoppers generally predominate over other kinds.

Game birds are eaten by the red fox when the opportunity arises, but these birds are more difficult to obtain than most of the foods in the animal's diet.

Moles, shrews, and weasels often are killed by the red fox but are left uneaten on the trails. There seems to be a particular lack of appetite for weasels.

Signs.—Prints of the feet are oval in shape, fig. 26; those of the front feet are about 2 inches long and 1¾ inches wide, and those of the hind feet are slightly smaller. On soft surfaces especially, prints of the front feet show a noticeable spread between the toes. The relatively slender and barlike impressions of the ball pads lie behind rather than between the toe marks. On mud and wet snow that are fine enough to hold the impressions, the furry nature of the feet may be detected. On such surfaces, footprints of the red fox and the gray fox may be told apart; a slender ball print is the mark of the red fox and a broader one the mark of the gray.

Track patterns made by the red fox at different gaits are similar to those made by the domestic dog, except that footprints made by a walking or trotting fox are usually in almost a straight line, not staggered; prints of hind and front feet on the same side of the fox generally register well. At both walking and trotting gaits, the red fox makes tracks that vary from 9 to 16 inches apart. Strides of 14 to 16 inches usually indicate a trot. Footprints of a galloping fox appear in groups of four; the distance between groups on level ground measures usually 30 to 88 inches. The length of each group of four prints and the distance between the groups increase with the rate of speed.

Scats of adult foxes range from short fragments to 12 inches in length, with an average length of 6.3 to 7.4 inches. Generally there are two to four segments to a passage and there may be as many as nine.

The den of the red fox is generally made from the burrow of a woodchuck, badger, or other animal; the burrow is enlarged to a diameter of 9 to 12 inches. Frequently it is in a dry, sloping bank or hillside with more than a 10 per cent grade. If in a comparatively level area, it is almost invariably on a slight elevation. Pathways, fig. 13, may radiate from the den, which is usually surrounded with scats, tufts of rabbit fur, feathers, bones, and the bodies of mice and other prey. Rabbit skulls left by foxes characteristically have the basal and nasal parts gnawed off, and only the central parts of the skulls remain.

Distribution.—The red fox is common the length and breadth of Illinois and is cyclic in abundance. Red foxes in this state are currently referred to the subspecies *Vulpes fulva fulva* (Desmarest), but further study may show that they are assignable to *V. f. regalis* Merriam. The range of the species includes most of Canada, all of Alaska, and much of the United States. It does not include parts of the southeastern and Gulf Coast states and certain regions bordering the Rocky Mountains and the Sierra Nevada.

UROCYON CINEREOARGENTEUS (Schreber)
Gray Fox

Description.—The gray fox, fig. 70, is slightly smaller and lighter in weight than the red fox. The upper parts are gray; the tail is black at the tip and along the upper side; the legs, feet, sides of neck, and back of each ear are rusty yellow; the throat and belly are whitish, bordered by a rusty or rufous color that continues onto the under side of the tail.

Length measurements: head and body 28–29 inches (720–750 mm.), tail 11–12 inches (280–300 mm.), over-all 39–41 inches (1,000–1,050 mm.), hind foot 4¾–5¾ inches (120–145 mm.), ear from notch 2⅜–3 inches (60–75 mm.). Weight: 5–12 pounds, average about 9½ pounds.

The skull is about 130 mm. (5⅛ inches) long and has two prominent ridges on each side of the braincase, fig. 60*b*; each

Fig. 70.—Gray fox.

lower jaw has an extra notch or step, fig. 60*d*, at the posterior end. Dental formula: I 3/3, C 1/1, Pm 4/4, M 2/3.

Life History.—The gray fox is an animal of the forest, fig. 1, and of river bottoms and bluffs, but it is sometimes seen in semiopen brushland. A reduction in lumbering and the establishment of forest preserves and other woodland sanctuaries in Illinois have provided ideal habitat for this intriguing species. The gray fox uses a ground den less frequently than does the red fox, and instead may choose a hollow tree, a hollow log, or a hole among some rocky cliffs. When pursued or startled, it may climb a tree, not as nimbly as a squirrel but with surprising speed, and take refuge among the small branches.

The gray fox appears to be even more omnivorous than the red fox and feeds on a variety of things, including quantities of berries, fruits, acorns, and seeds, as well as mice, rabbits, insects, and, occasionally, birds.

Three to five young per litter are born in the spring. In development and behavior they are very similar to the young of the red fox.

Signs.—Tracks of the gray fox, fig. 27, can be distinguished from those of the red fox only when they are made under the most ideal conditions. Impressions of the relatively thick ball pads of the gray fox may sometimes be seen in thin layers of mud or snow.

The droppings of the gray fox, like those of the red, contain large amounts of rabbit and mouse fur and bones, and, during the summer and fall, usually fruit seeds.

Distribution.—The gray fox is common in Illinois only in rather heavily wooded areas, but is state-wide in distribution. It is represented in Illinois by the subspecies *Urocyon cinereoargenteus cinereoargenteus* (Schreber). The range of the species includes the eastern half of the United States and most of the southwestern quarter; also it includes Mexico and countries southward as far as Colombia. It does not include much of the short-grass prairie area or the Rocky Mountain region north of Colorado.

CANIS LATRANS Say

Coyote   Prairie Wolf

Description.—The coyote, fig. 71, somewhat resembles a German "shepherd" (police) dog in size, conformation, and color, but it carries its tail below the level of its back rather than curved upward. The upper part of the body is a grizzled gray or buff, the muzzle reddish brown or gray, and the under parts are whitish, cream colored, or pinkish yellow. The tail is bushy, the snout narrow and long, the ears are pointed, and the legs long.

Length measurements: head and body 32–39 inches (815–990 mm.); tail about 12–15 inches (300–375 mm.); over-all 44–54 inches (1,120–1,375 mm.); hind foot about 7–8 inches (180–211 mm.); ear about 4–4½ inches (100–115 mm.). Weight: 25 pounds to possibly 55 pounds.

The skull of the coyote is 180–200 mm. (7–8 inches) long; its zygomatic breadth is 97–110 mm. (3¾–4¼ inches). Dental formula: I 3/3, C 1/1, Pm 4/4, M 2/3.

Coyotes and dogs may cross or hybridize. Several hybrids of these animals are known from Illinois. In size, markings, and bearing, they naturally are somewhat intermediate between the coyote and dog parents. It is possible to distinguish skulls of some coyotes, dogs, and first generation coyote-dog hybrids by taking certain cranial measurements and determining cranial indices. Obviously, skulls with intermediate or near-intermediate measurements are from animals having both dog and coyote ancestry. Convenient indices and formulas for deriving them are shown below:

| | | CRANIAL INDEX | | |
FORMULA		Coyote	Hybrid	Dog
$\dfrac{\text{Palatal width between inner margins of alveoli of upper first premolars}}{\text{Alveolar length of upper premolars and molars}} \times 100$		25–32	32–38	32–52
$\dfrac{\text{Width of basioccipital bone}}{\text{Width of braincase}} \times 100$		23.3–28.1	28.0–29.1	29.1–34.3
$\dfrac{\text{Depth of lower jaw below second molar}}{\text{Length of lower jaw}} \times 100$		12.2–14.1	14.2–14.7	14.0–17.8

The coyote of Illinois can be distinguished from the timber wolf of Minnesota or Michigan by its smaller size (less than 62 inches over-all length not including hair on tip of tail), narrower nose pad (less than 1⅛ inches in width), smaller feet (each hind foot less than 10 inches long), smaller skull (length less than 8¼ inches), and smaller teeth. The timber or gray wolf does not now occur in Illinois.

Life History.—Although the coyote may never have been abundant in Illinois, it does not seem to have suffered much from the settlement of the land by man. In fact, it may have profited by the opening up of the land. Over a recent 5-year period, 65 coyotes were taken in Fulton County, central Illinois, according to a report by Anderson (1951:172). Since coyotes are interbreeding with dogs, and several hybrids have already been taken in Illinois, it will be interesting to determine if the number of crosses increases.

Probably the coyote in Illinois feeds princ'pally on rabbits and rodents, as is the case westward, but no one has studied its food habits in this state. It feeds also on insects, vegetable matter, birds, and carrion. Some individuals kill poultry and livestock. A coyote-dog hybrid from along the Illinois River south of Hennepin had mice and traces of rabbit in its stomach.

In Illinois, as elsewhere in central United States, the female coyote has probably one litter each year, with an average of about half a dozen young. The young are whelped in a den and spend their first few weeks there. As soon as they are old enough to eat solid food, they are fed by both parents. When 3 or 4 weeks old the pups play at the entrance to the den and at 10 weeks of age they may abandon the den completely.

The coyote, known also as the prairie wolf or brush wolf, is an inhabitant of open country rather than woods. An early writer on Illinois (Anonymous 1837:40) commented, "The prairie wolf . . . takes its name from its habit of residing entirely upon

Fig. 71.—Coyote.

the open plains. . . . The most friendly relations subsist between it and the common wolf, and they constantly hunt in packs together. Nothing is more common than to see a large black wolf in company with several prairie wolves." The actual coyote-wolf relationship is less friendly than indicated above. Coyotes often trail wolves for kills left by the larger animals.

Signs.—Coyote tracks, fig. 24, cannot be distinguished with certainty from some dog tracks. The print of a hind foot of the coyote commonly measures about 2¾ inches long, while that of a front foot is slightly shorter. The register of the tracks of a walking animal is not perfect; the toe marks of each hind foot fall at about the center of the track of the front foot on the same side of the body.

The den of the coyote is usually in a bank or hillside but sometimes in level ground. It commonly has only one entrance, which has a mound of earth in front. Well-worn paths radiate from the entrance and almost invariably lead to one side of the mound rather than over it.

Distribution.—The coyote occurs in much of Illinois, but it is not common anywhere in the state. Individuals in northern and central Illinois belong to the subspecies *Canis latrans thamnos* Jackson; those in southern Illinois are presumably referable to *C. l. frustror* Woodhouse. The range of the species includes most of western North America (central Alaska almost to Panama) and an eastward extension that continues through most of the Great Lakes region and ends in a narrow area north of the St. Lawrence River.

CANIS LUPUS (Linnaeus)

Timber Wolf   Gray Wolf

Description.—The timber wolf is a large, long-haired animal weighing 75 pounds or more and attaining a length of nearly 5½ feet. The color is usually gray or light gray-brown. The muzzle is heavy and blunt, and the nose pad exceeds 1⅛ inches in diameter. The ears are less pointed than those of the coyote.

The skull is about 255 mm. (10 inches) long, and the teeth are large (half again as large as the corresponding teeth in the coyote) and arranged as in the coyote.

Distribution.—In the late 1700's and early 1800's, the timber wolf was common in the country that is now Illinois, according

to accounts of early explorers and settlers. Reports indicate that
it attacked large game animals and livestock, even hauling down
young horses. The wolf probably fed rather extensively on
bison, elk, and deer before the coming of the white man to the
Illinois country.

By the middle 1800's, the timber wolf in Illinois had been
greatly reduced in numbers and before the end of the century
it had been completely eliminated from the state. The wolf
bounties that have been paid since that time are for coyotes or
prairie wolves, dogs, or coyote-dog hybrids. The subspecies that
occurred in Illinois was *Canis lupus lycaon* Schreber. The species has a range that includes most of Canada and Alaska and
parts of some north-central states, including Wisconsin. Also,
it occurs in isolated areas of Oregon, Colorado, Utah, Arizona,
New Mexico, and Mexico.

CANIS NIGER Bartram
Red Wolf

Description.—The red wolf is intermediate in size between
the timber wolf and the coyote. It is about 5 feet long and
weighs 60 to 70 pounds. The broad nose and the rounded ears
are more like those of the timber wolf than those of the coyote.
The sides of the muzzle are buffy red and the coat color has a
buffy cast.

The skull is slightly smaller (length about 245 mm. or 9⅝
inches) than that of the timber wolf; the crowns of the teeth
are more deeply cleft, and the shearing edges more bladelike.
The tooth arrangement is the same as that of the coyote and
of the timber wolf.

Distribution.—In former times the red wolf occurred probably along the Mississippi and other rivers in the southern part
of Illinois. A young male, now preserved in the American Museum of Natural History, was obtained on February 7, 1893,
at Warsaw, Illinois, by C. K. Worthen. The red wolf still
occurs in fair numbers in the Ozark Mountains of Missouri
and Arkansas and southward to Louisiana and eastern Texas.
There is the possibility that some individuals may live in, or
stray into, the heavily wooded portions of extreme southern
Illinois. The subspecies in Illinois was *Canis niger gregoryi*
Goldman.

FELIS CONCOLOR Linnaeus

Cougar   Mountain Lion   Panther

Description.—The cougar is a large, long-tailed cat measuring nearly 80 inches from the nose to the tip of the tail. The body is light brown or tawny; the backs of the ears and the tip of the tail are dark brown.

The large skull is about 8 inches long, and the cheek teeth are bladelike. Dental formula: I 3/3, C 1/1, Pm 3/2, M 1/1.

Distribution.—Accounts of early settlers and contemporary travelers indicate that in the first half of the 1800's cougars or panthers were found in the forested parts of Illinois. After the middle of the century, the cougar (rarely called mountain lion in Illinois) disappeared very rapidly. The form present was *Felis concolor couguar* Kerr. The range of the species now includes part of northern British Columbia and an extensive area southward through the United States (principally west of the Rocky Mountains) and into South America, along the Gulf Coast from Mexico to western Florida, and peninsular Florida.

LYNX RUFUS (Schreber)

Bobcat

Description.—The bobcat, fig. 72, is a short-tailed cat standing 20 to 23 inches high at the shoulder. Its body is mostly yellowish gray, with a sprinkling of black; the under parts and inner surfaces of the legs are whitish, spotted with black; streaks on the long fur covering the cheeks are dark gray; and the upper part of the tip of the tail is black. Each pointed ear bears a small tuft or pencil of hairs.

Length measurements: head and body about 27 inches (625–755 mm.), tail about 5¼ inches (135 mm.), over-all about 33 inches (760–880 mm.), hind foot about 7 inches (170–188 mm.). Weight: 15–25 pounds, usually about 20 pounds.

The skull is 105–135 mm. (about 5 inches) long. Except for having three rather than four cheek teeth (premolars and molar) in each half of the upper jaw, it is not unlike that of the house cat. Dental formula: I 3/3, C 1/1, Pm 2/2, M 1/1.

Life History.—The bobcat frequents wooded sections along rivers, especially timbered bluffs and slopes that are interspersed with sunny glades and swampy bottomlands. Broken country

Fig. 72.—Bobcat.

provides an ideal hunting ground for the bobcat, since mice, rabbits, squirrels, birds, and insects upon which it feeds abound there.

A bobcat den may be located under a log, within a fallen hollow log, or within a standing hollow tree. A nest in a dense thicket may occasionally serve as a home. Little is known about the life history of the bobcat in Illinois. Information gained from studies in other states indicates that young are born in March or April, that there is one litter of about three each year, and that the young leave the care of their mother and the company of their brothers and sisters late in the summer. The rate of development of the young bobcat is similar to that of the young of the domestic cat.

The bobcat apparently covers a large territory in its hunting, for an individual seen in a given place one day may be a considerable distance away the next week. This cat, especially during the mating season, may render a weird and eerie series of yowls and meows somewhat like that of the common house cat, but louder, huskier, and infinitely more mysterious. A harsh,

mysterious scream has led persons to believe a lion or "cata-
mount" to be present, when only a bobcat has been hunting in
the neighborhood.

Signs.—The bobcat makes tracks, fig. 19, similar in shape to
those of the house cat, but the prints of the individual feet are
each about 2 inches in diameter (as compared to 1¼ inches for
those of the domestic cat), and the heel pads produce more complicated patterns. When the animal is walking, the hind feet
(which are smaller than the front feet) often overstep the
front feet.

Distribution.—The bobcat, which is now rare in Illinois,
occurs in wooded bottomlands, fig. 3, of some of the major
rivers in heavily wooded regions of southern and possibly of
northwestern Illinois. The Illinois bobcat belongs to the subspecies *Lynx rufus rufus* (Schreber). The present range of the
species includes western North America from southern Canada
to central Mexico; across northern United States and southern
Canada to Nova Scotia and the lower Appalachians; across
southern United States to southern South Carolina.

ORDER RODENTIA
Rodents or Gnawing Animals

The rodents, or gnawing animals, have only two incisors or
front teeth and no canine teeth in each jaw. The incisors of the
upper jaw are especially large and chisel shaped. Most rodents
occurring wild in Illinois have relatively long tails. In both
numbers of kinds and of individuals, the rodents are the most
numerous of all mammals. They comprise about three-fifths of
the number of species of the North American mammalian fauna.
Twenty-five species of rodents are found in Illinois, most of
which are various kinds of squirrels, rats, or mice. The largest
rodent in Illinois is the beaver; the smallest is the western harvest mouse.

Economic Status.—Members of the order Rodentia, both
native and introduced, are of tremendous economic importance
in Illinois. For example, a single species, the Norway rat, costs
the taxpayers several million dollars a year as a result of the
spoilage of food and grain, damage to buildings and merchandise,
predation on poultry and native birds, and transmission of disease. Money spent in attempts to control this rat amounts to

millions more. Its small cousin, the house mouse, is less destructive individually, but because of its greater abundance causes almost as much total damage. Both the Norway rat and the house mouse were introduced from Europe.

Some of the native rodents of Illinois are responsible for considerable economic loss. Some have direct or indirect economic value. Woodchucks, pocket gophers, ground squirrels, native mice, muskrats, beavers, and chipmunks do varying amounts of damage to gardens, cereal grains, and silage crops. Field mice and voles cause considerable loss in orchards as a result of their girdling of trees, fig. 37. By their burrowing, such animals as muskrats and woodchucks weaken dams and embankments. Pocket gophers and woodchucks make mounds of earth that interfere with mowing operations. Beavers construct dams that sometimes cause flooding of farmland. Some kinds of squirrels occasionally eat young birds and bird eggs; some may be nuisances in human habitations when they store food or make their homes in attics or walls. A few native mice feed on clover and grasses and they may also deface lawns with their runways.

Among the native rodents of direct or indirect value are the ground-inhabiting squirrels and mice, which play a role in soil improvement by bringing subsoil to the surface and depositing humus-forming material below the surface of the ground. Their burrows hold excess water during wet periods and thus serve as underground reservoirs during dry periods. These burrows are frequently used as nests by bumblebees, which help pollinate the blossoms of some crop plants. Tree squirrels and chipmunks contribute to forest production by burying tree seeds in the ground and failing to return for them; some of these buried seeds sprout and grow. Fox squirrels and gray squirrels provide food and sport for countless hunters and furnish aesthetic enjoyment for nearly everyone. Some mice feed on insects and may aid in the control of insect pests more than is realized. Most species of rodents serve as food for valuable fur-bearing mammals. Woodchucks make and desert numerous burrows that are used by furbearers as dens for rearing their young and by cottontails as shelter from harsh weather.

The muskrat is the most important fur resource of Illinois. Each year an estimated 20,000 people in Illinois harvest muskrats; the annual value of the furs they take is somewhat under a million dollars.

KEY TO SPECIES

Whole Animals

1. Hair at mid-length of tail longer than diameter of tail.... 2
   Hair at mid-length of tail shorter than diameter of tail, or
   hair absent...... 9
2. Tail not more than ¼ over-all length of animal; feet black;
   skull more than 70 mm. (2¾ in.) long...............
   ....woodchuck, *Marmota monax*
   Tail more than ¼ over-all length of animal; feet not
   black; skull less than 70 mm. long.................... 3
3. A large, loose fold of skin between front and hind legs,
   fig. 83; bush of tail in cross section decidedly flattened..
   ....southern flying squirrel, *Glaucomys volans*
   No loose fold of skin between legs; bush of tail round or
   oval in cross section................................. 4
4. Hair at mid-length of tail less than 20 mm. (¾ in.) long.. 5
   Hair at mid-length of tail more than 35 mm. (1⅜ in.) long. 7
5. Body without longitudinal stripes, fig. 78...............
   ....Franklin's ground squirrel, *Citellus franklinii*
   Body with longitudinal stripes........................ 6
6. Body with 2 longitudinal light stripes bordered with dark
   brown, these stripes represented also on the sides of the
   face, fig. 79.............eastern chipmunk, *Tamias striatus*
   Body with more than 2 longitudinal light stripes bordered
   with dark brown, no stripes present on the sides of the
   face, fig. 76...
   ....thirteen-lined ground squirrel, *Citellus tridecemlineatus*
7. Over-all length of animal less than 350 mm. (13¾ in.); a
   short black stripe bordering the belly color on each side,
   fig. 80..............red squirrel, *Tamiasciurus hudsonicus*†
   Over-all length of animal more than 350 mm.; no lateral
   dark stripe bordering belly color..................... 8
8. Upper parts of animal uniform gray, under parts gray or
   white; tail with white-tipped hairs...................
   ....eastern gray squirrel, *Sciurus carolinensis*
   Upper parts of animal dappled gray, under parts reddish;
   tail with red-tipped hairs............................
   ....eastern fox squirrel, *Sciurus niger*
9. Claws on front feet, fig. 85, 3 times as large as claws on hind
   feet; each cheek with a pouch that opens by an external,
   curved slit starting near the side of the mouth.........
   ....plains pocket gopher, *Geomys bursarius*
   Claws on front and hind feet subequal in size; no external
   cheek pouches 10
10. Tail flattened dorsoventrally, fig. 87, more than 50 mm.
    (2 in.) wide; head+body more than 400 mm. (15¾ in.)
    long...........................beaver, *Castor canadensis*

---

†This species probably is not now present in Illinois.

Tail not flattened dorsoventrally, and less than 25 mm. (1
  in.) wide; head+body less than 400 mm. long......... 11
11. Tail at least 1⅓ times the length of head+body, fig. 102;
  upper jaw with 4 cheek teeth on each side.............
  ....meadow jumping mouse, *Zapus hudsonius*
  Tail less than 1⅓ times the length of head+body; upper
  jaw with 3 cheek teeth on each side................... 12
12. Tail flattened laterally; front feet each with 5 clawed toes
  ....muskrat, *Ondatra zibethicus*
  Tail not flattened; front feet each with 4 clawed toes..... 13
13. Tail less than ⅓ length of head+body.................. 14
  Tail more than ⅓ length of head+body................. 17
14. Upper incisors each with shallow longitudinal groove on
  face, fig. 74*i*.....southern bog lemming, *Synaptomys cooperi*
  Upper incisors not grooved............................ 15
15. Tail less than 25 mm. (1 in.) long; hind foot less than 18
  mm. (almost ¾ in.) long; ears hidden in soft fur........
  ....pine vole, *Pitymys pinetorum*
  Tail more than 25 mm. long; hind foot usually more than
  18 mm. long; ears projecting above coarse fur 16
16. Fur of under parts tipped with white; feet black; each last
  upper molar with 5 or 6 enamel loops, fig. 74*d*...........
  ....meadow vole, *Microtus pennsylvanicus*
  Fur of under parts ochraceous; feet brown; each last upper molar with only 4 enamel loops, fig. 74*e*...........
  ....prairie vole, *Microtus ochrogaster*
17. Upper incisors each with a longitudinal groove........... 18
  Upper incisors not grooved 19
18. Body having sides washed with fulvous; back brown.......
  ....western harvest mouse, *Reithrodontomys megalotis*
  Body having sides and back rich brown.................
  ....eastern harvest mouse, *Reithrodontomys humilis**
19. Over-all length of animal more than 200 mm. (8 in.)..... 20
  Over-all length of animal less than 200 mm............. 24
20. Fur on throat with each hair white its entire length; upper
  cheek teeth with triangles but without distinct cusps,
  fig. 74*b*..............eastern wood rat, *Neotoma floridana*
  Fur on throat with each hair white distally but darkened at
  base; upper cheek teeth with distinct cusps rather than
  triangles, fig. 74*a, c*................................ 21
21. Hind foot not more than 34 mm. (1⅜ in.) long; under side
  of tail light; cusps on upper cheek teeth in 2 rows...... 22
  Hind foot more than 34 mm. long; under side and upper
  side of tail gray; cusps on upper cheek teeth in 3 rows.. 23
22. Fur long, grizzled buff and black; ears almost hidden in fur
  ....hispid cotton rat, *Sigmodon hispidus**
  Fur short, brown above and not grizzled; ears projecting
  above fur...................rice rat, *Oryzomys palustris*
23. Tail shorter than length of head+body.................
  ....Norway rat, *Rattus norvegicus*

*This species may occur in Illinois, but there are no official records of it here.

Tail as long as or longer than head+body...............
..................................roof rat, *Rattus rattus*
24. Belly and feet dirty gray or olive..house mouse, *Mus musculus*
Belly and feet white.. 25
25. Head, ears, back, and sides of a uniform reddish color....
........................golden mouse, *Peromyscus nuttalli*
Head, ears, and back brown or black; sides variable and
lighter in color.. 26
26. Hind foot 18 mm. (¾ in.) or less in length; tail usually
less than 60 mm. (2⅜ in.) in length; middorsal dark
stripe prominent.......deer mouse, *Peromyscus maniculatus*
Hind foot more than 18 mm. in length; tail usually more
than 60 mm. in length; middorsal dark stripe present or
absent 27
27. Hind foot usually less than 23 mm. (⅞ in.) in length;
head+body usually less than 100 mm. (4 in.) in length.....
...................white-footed mouse, *Peromyscus leucopus*
Hind foot usually 23 mm. or more in length; head+body
usually more than 100 mm. in length...................
....................cotton mouse, *Peromyscus gossypinus*

Skulls

1. Each frontal bone with well-developed postorbital process,
figs. 73*a, b, e,* 74*j, k*................................. 2
Each frontal bone without postorbital process............. 9
2. Postorbital processes large, projecting at right angles to the
longitudinal axis of the skull, fig. 73*e*; incisors whitish
..........................woodchuck, *Marmota monax*
Postorbital processes small and angled backward; incisors
yellowish 3
3. Interorbital region deeply notched anterior to each postorbital process, fig. 74*j* .
...........southern flying squirrel, *Glaucomys volans*
Interorbital region not deeply notched anterior to each postorbital process, figs. 73*a*, 74*k*.......................... 4
4. Zygomatic arches tending to converge anteriorly, fig. 73*a*.. 5
Zygomatic arches nearly parallel, fig. 74*k*................. 7
5. Nasals more than 17 mm. (⅝ in.) long...................
............Franklin's ground squirrel, *Citellus franklinii*
Nasals less than 17 mm. long.......................... 6
6. Upper jaw with 5 cheek teeth on each side...............
....thirteen-lined ground squirrel, *Citellus tridecemlineatus*
Upper jaw with 4 cheek teeth on each side..............
......................eastern chipmunk, *Tamias striatus*
7. Upper jaw with series of cheek teeth on each side less
than 10 mm. (⅜ in.) long..........................
..................red squirrel, *Tamiasciurus hudsonicus*†
Upper jaw with series of cheek teeth on each side more than
10 mm. long.. 8

†This species probably is not now present in Illinois.

8. Upper jaw with 5 cheek teeth on each side; anteriormost
   tooth rudimentary..eastern gray squirrel, *Sciurus carolinensis*
   Upper jaw with 4 cheek teeth on each side...............
   ....eastern fox squirrel, *Sciurus niger*
9. Upper incisors each with 2 longitudinal grooves, fig. 73*c*..
   ....plains pocket gopher, *Geomys bursarius*
   Upper incisors each with 1 or no groove................. 10

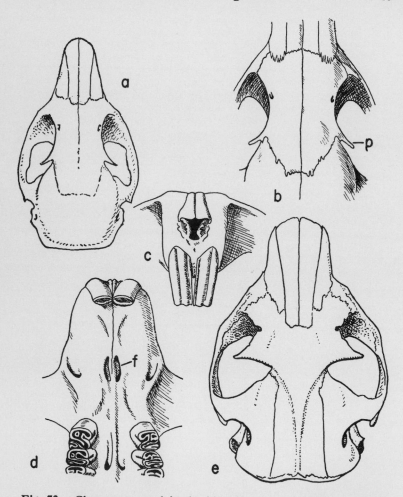

Fig. 73.—Characters used in the identification of rodents: *a,* skull
of thirteen-lined ground squirrel, top view; *b,* portion of skull of
Franklin's ground squirrel, top view (*p,* postorbital process); *c,*
front of rostrum of pocket gopher; *d,* rostrum of beaver, under
side (*f,* incisive foramen); *e,* skull of woodchuck, top view.

10. Skull length more than 90 mm. (3½ in.) ; length of incisive
    foramen, fig. 73*d*, shorter than length of grinding sur-
    faces of first two upper cheek teeth beaver, *Castor canadensis*
    Skull length less than 75 mm. (3 in.) ; length of incisive
    foramen greater than length of grinding surfaces of first
    2 upper cheek teeth................................... 11
11. Upper jaw with 4 cheek teeth on each side...............
    ....meadow jumping mouse, *Zapus hudsonius*
    Upper jaw with 3 cheek teeth on each side.............. 12
12. Skull length more than 45 mm. (1¾ in.)................
    ....muskrat, *Ondatra zibethicus*
    Skull length less than 45 mm........................... 13
13. Skull length 30 mm. (1⅛ in.) or less.................... 14
    Skull length more than 30 mm......................... 23

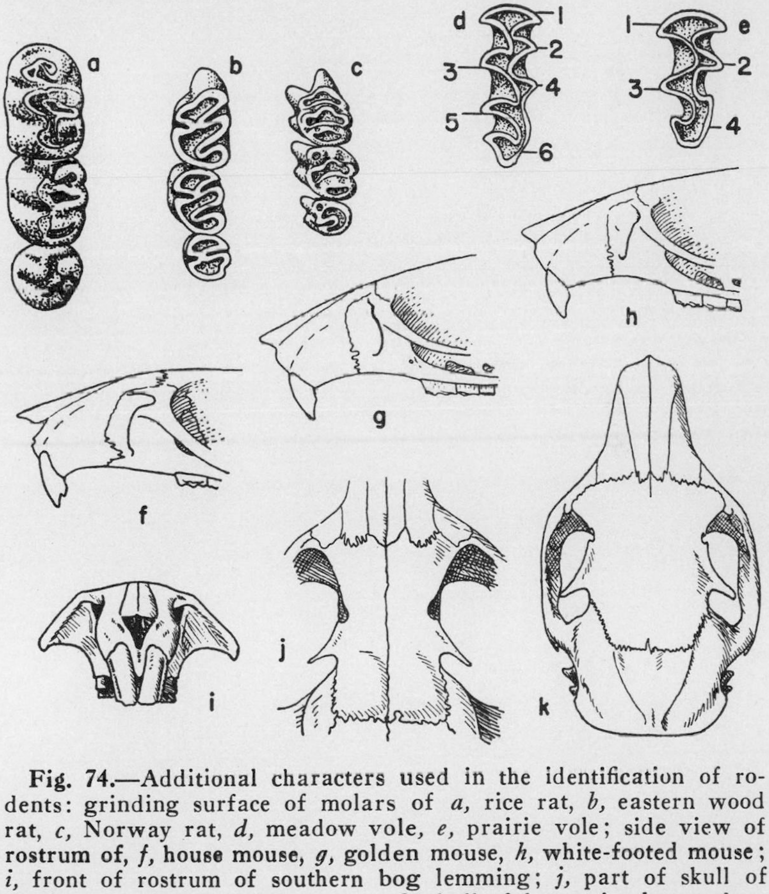

Fig. 74.—Additional characters used in the identification of rodents: grinding surface of molars of *a*, rice rat, *b*, eastern wood
rat, *c*, Norway rat, *d*, meadow vole, *e*, prairie vole; side view of
rostrum of, *f*, house mouse, *g*, golden mouse, *h*, white-footed mouse ;
i, front of rostrum of southern bog lemming; *j*, part of skull of
southern flying squirrel, top view; *k*, skull of fox squirrel, top view.

14. Upper molars with crowns having rows of tubercules or
    cusps, fig. 74a, c... 15
    Upper molars with crowns having enamel loops, fig. 74d, e,
    rather than tubercules or cusps......................... 20
15. Upper incisors with longitudinal groove on front surface or
    face of each............harvest mice, *Reithrodontomys* spp.
    Upper incisors without grooves on faces................. 16
16. Upper incisors, viewed from side, each with a pronounced
    terminal notch, fig. 74f..........house mouse, *Mus musculus*
    Upper incisors, viewed from side, without terminal notches 17
17. Infraorbital plate with anterior margin straight, fig. 74g
    ....golden mouse, *Peromyscus nuttalli*
    Infraorbital plate with anterior margin bowed forward, as
    in fig. 74h... 18
18. Zygomatic width of skull anteriorly less than 12 mm. (½
    in.) ....deer mouse, *Peromyscus maniculatus*
    Zygomatic width of skull anteriorly more than 12 mm...... 19
19. Maxillary tooth rows each 4 mm. long or less.............
    ....white-footed mouse, *Peromyscus leucopus*
    Maxillary tooth rows each usually more than 4 mm. long..
    .... cotton mouse, *Peromyscus gossypinus*
20. Upper incisors each with longitudinal groove on face or
    front surface, fig. 74i.................................
    ....southern bog lemming, *Synaptomys cooperi*
    Upper incisors without grooves on faces.................. 21
21. Interorbital region more than 4 mm. wide...............
    ....pine vole, *Pitymys pinetorum*
    Interorbital region less than 4 mm. wide................ 22
22. Last upper molars each with 5 or 6 enamel loops, fig. 74d..
    .... meadow vole, *Microtus pennsylvanicus*
    Last upper molars each with 4 enamel loops, fig. 74e.......
    ....prairie vole, *Microtus ochrogaster*
23. Upper molars with irregular triangles on crowns, fig. 74b..
    ....eastern wood rat, *Neotoma floridana*
    Upper molars with rows of cusps on crowns, fig. 74a, c..... 24
24. Upper cheek teeth with cusps in 2 rows; skull less than
    35 mm. (1⅜ in.) long................................ 25
    Upper cheek teeth with cusps in 3 rows; skull more than
    35 mm. long... 26
25. Anterior half of rostrum wider than least interorbital space
    ....hispid cotton rat, *Sigmodon hispidus**
    Anterior half of rostrum narrower than least interorbital
    space.......................rice rat, *Oryzomys palustris*
26. Length of each parietal, measured along temporal ridge,
    less than greatest distance between temporal ridges.....
    ....roof rat, *Rattus rattus*
    Length of each parietal, measured along temporal ridge,
    about equal to distance between temporal ridges.........
    ....Norway rat, *Rattus norvegicus*

*This species may occur in Illinois, but there are no official records of it here.

MARMOTA MONAX (Linnaeus)
Woodchuck   Groundhog

Description.—The woodchuck, fig. 75, is a stout-bodied, squirrel-like mammal with short, powerful legs, short ears, short blunt muzzle, and relatively short furry tail. In most individuals the upper parts are a grizzled yellowish gray to dark brown; the under parts are lighter, often with a reddish wash. The feet are dark brown or black. Occasional woodchucks are almost black. The four toes of each front foot and the five toes of each hind foot have well-developed claws.

Length measurements: head and body $16\frac{1}{4}$–$19\frac{1}{2}$ inches (415–495 mm.); tail $4\frac{1}{4}$–6 inches (110–155 mm.); over-all $20\frac{1}{2}$–$25\frac{1}{2}$ inches (525–650 mm.); hind foot 3–$3\frac{3}{4}$ inches (75–95 mm.). Weight: average about $6\frac{1}{2}$ pounds, that of the male slightly greater than that of the female.

The flat skull, fig. 73e, has well-developed postorbital processes and is 75–100 mm. (3–4 inches) long. The strong incisors are cream colored or white. Dental formula: I 1/1, C 0/0, Pm 2/1, M 3/3.

Life History.—The woodchuck is normally an animal of the forest edge. It prefers rolling land that is well drained, fig. 2. In some parts of Illinois, the woodchuck lives along wooded river bluffs, in other parts in more open country, and in a few areas in heavy woods. In level countryside, it may make its burrow in a railroad or highway embankment, fig. 5.

As with most other squirrel-like animals, the woodchuck restricts its activity to daylight hours. It digs innumerable burrows, most of which at some time become the homes of other mammals. At one of the enlarged entranceways to a burrow it may be seen dozing in the warm sun, or busily trying to rid itself of fleas, lice, or ticks, or sitting upright to get a good view of its surroundings. It may occasionally climb a tree.

The female is occupied during the summer months in rearing a family. The nest chamber is in the burrow and so situated as to remain dry. The nest itself consists of only a few leaves or blades of grass. Two to six (usually three or four) young per litter are born in April or May. The young, fig. 75, grow and mature rapidly, and before long are busily feeding on green plants. Both young and adults commonly feed on clover, alfalfa, dandelions, wild lettuce, and plantain.

Fig. 75.—Woodchucks: top, young; bottom, adult.

By late October or thereabouts, when most native herbs and grasses are dry, when the amount of daylight has greatly decreased, and the days as well as the nights are frosty, the woodchuck gradually becomes less active above ground and is soon no longer seen. In its burrow, it sinks into the deep sleep of true hibernation, in which the heart beat and respiratory rates are only one-tenth as fast as in normal life. Should a warm spell occur, the woodchuck may arouse and leave its hibernating chamber for a brief time. Indeed, the groundhog, as the woodchuck is sometimes called, may even be above ground for a short time on the second day of February, but not motivated by any desire to view its shadow.

Signs.—During the growing seasons of the year, woodchuck tracks, fig. 29, may often be found along creek beds or on dusty or muddy roads in woods. The four-toed prints of the front feet and the five-toed prints of the hind feet are closely bunched when the woodchuck runs; the distance between each set is normally 12 inches. About 4 inches separate the sets of front and hind prints when the chuck is walking. Each individual footprint is about $1\frac{1}{2}$ inches long.

A woodchuck burrow is usually 6 to 12 inches wide at each of its two or three entrances but soon narrows inside and at a depth of 2 feet may be only 4 to 7 inches in diameter. Because the woodchuck keeps its burrow clean by constant enlarging, there is usually a fresh mound of loose dirt near one or another of the openings. A burrow deserted by a woodchuck and occupied by a skunk or a rabbit does not have such a fresh mound. Radiating from the entranceway of a woodchuck burrow are partially concealed paths or runways which serve as avenues to food supplies. Closely cropped herbage near the entrance is an indication that a woodchuck occupies the burrow.

Distribution.—The woodchuck, common in many parts of Illinois, occurs the length and breadth of the state, with the possible exception of some parts of the level black soil regions of central Illinois. Only one subspecies, *Marmota monax monax* (Linnaeus), is known in this state. The species ranges from Labrador westward to central British Columbia, then northward through Yukon to eastern Alaska; southward as far as northern Idaho in the western part of the United States and as far as northeastern Oklahoma and northern Georgia and Alabama in the central and eastern parts.

Fig. 76.—Thirteen-lined ground squirrel.

CITELLUS TRIDECEMLINEATUS (Mitchill)

Thirteen-Lined Ground Squirrel   Striped Gopher

Description.—The thirteen-lined ground squirrel, fig. 76, is about the size of a chipmunk. The ears are small, the eyes are large, and the slightly bushy tail is shorter than the body. The back has approximately 13 longitudinal stripes alternately dark brown and buff. Each of the brown stripes contains a series of buff spots. The sides of the body are yellowish or buff, and the belly is the same color as the light stripes on the back.

Length measurements: head and body 5¾–7¾ inches (145–200 mm.); tail 3⅛–4¼ inches (80–110 mm.); over-all 8¾–12¼ inches (225–310 mm.); hind foot 1¼–1⅝ inches (32–41 mm.). Weight: about 4 ounces in early summer; nearly 8 ounces in fall prior to hibernation.

The skull, fig. 73a, is 36–43 mm. (about 1½ inches) long and delicate in appearance. In Illinois, only this squirrel, the gray squirrel, woodchuck, Franklin's ground squirrel, and the southern flying squirrel have five cheek teeth on each side of the upper jaw. The skull differs from that of the gray squirrel and the woodchuck by its decidedly smaller size; from that of Franklin's ground squirrel by its smaller size and by its shorter nasal bones (less than 17 mm. long); from that of the southern flying squirrel by the absence of deep notches in the interorbital region, fig. 73a, by incisors that have yellow rather than reddish front faces, and by smaller auditory bullae. Dental formula: I 1/1, C 0/0, Pm 2/1, M 3/3.

Life History.—In the northern half of Illinois, the thirteen-lined ground squirrel is seen through casual observations perhaps more often than any other wild mammal during summer days. Members of this species attract attention as they dash across highways or as they stand stiffly erect near the entrances of their burrows on golf courses, in pastures, in cemeteries, and on grassy roadsides. The shortness of the grass in areas where they prefer to live makes them conspicuous, fig. 2. Loud noises or sudden movements may cause them to run for their burrows. A ground squirrel may give its shrill, quavering whistle of alarm before disappearing underground.

So great is its curiosity that a thirteen-liner cannot long remain hidden. Soon it will be peering out of the mouth of its burrow and before much longer it will be back at its usual activities of feeding, gathering grasses for its nest, enlarging its burrow, cutting paths through the grass, or sunning itself. It is always alert for a weasel, badger, dog, house cat, man, or hawk.

The thirteen-lined ground squirrel feeds on seeds, on grasses and other herbs, and on insects. At times during the summer, nearly half its food may consist of grasshoppers, white grubs, webworms, cutworms, and other insects.

Young, numbering usually 7 to 10 per litter, are born in late April or early May in a grass-lined nest below the surface of the ground. By mid-June, the young have grown from naked, blind creatures to thin-bodied, big-headed, furred animals able to clamber out of the burrow and nibble on herbs. By mid-July, some young have dug simple burrows of their own.

As summer progresses, this ground squirrel lays on quantities of body fat, and by the time it enters hibernation its early summer weight has nearly doubled. In the fall when the days get shorter and colder, this squirrel spends more time below ground and becomes more and more sluggish. Usually by the time grasses and seeds have been covered by snow, the ground squirrel has sealed the burrow entrance from within and has retired to a hibernating cell, which is just large enough for the animal and a little nest material. The cell is to one side of the burrow proper and so situated as to remain dry. The animal rolls up in a ball, with its nose tucked against its belly near its hind legs, and goes into a deep sleep. The rates of breathing and heart beat are greatly reduced, and the animal's body temperature becomes nearly the same as that of its surroundings. When the animal hibernates, it cannot effectively control or regulate its temperature and therefore, if the temperature of the hibernating chamber drops below the freezing point, the animal will die.

If the weather becomes unseasonably warm in winter, the thirteen-lined ground squirrel may appear above ground for short periods. Otherwise it remains below ground until March or April. At this time it is thin, for its stored body fat has been used during the long period of hibernation.

Signs.—The print of each front foot shows only four toes; the print of each hind foot shows five toes; that of the hind foot measures about $1\frac{1}{2}$ inches. A burrow entrance is usually slightly less than 2 inches in diameter and without a pile of dirt at the opening.

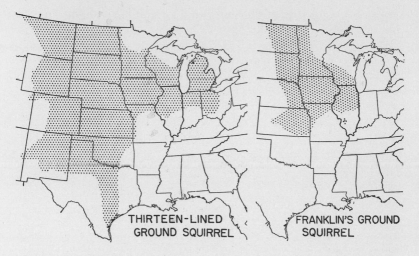

Fig. 77.—Known distribution, in the United States, of the Illinois ground squirrels.

Distribution.—The thirteen-lined ground squirrel is abundant in the northern two-thirds of Illinois, but is unknown south of an imaginary line connecting Clark and Madison counties. One subspecies, *Citellus tridecemlineatus tridecemlineatus* (Mitchill), occurs in this state. The range of the species extends from central Ohio and eastern Michigan northward and westward through Wisconsin and northern Illinois into south-central Canada, then southward to eastern Arizona, New Mexico, and south-central Texas, fig. 77.

CITELLUS FRANKLINII (Sabine)

Franklin's Ground Squirrel   Gray Gopher

Description.—Franklin's ground squirrel, fig. 78, is only slightly smaller than the arboreal gray squirrel, which it somewhat resembles. Its ears are short and its tail is bushy, but not so bushy or so long as that of the gray squirrel. The back is brownish gray, speckled with black, and has an overwash of yellow. The head and tail are gray, and the under parts are of a grayish color not much lighter than that of the back.

Length measurements: head and body 8½–9½ inches (220–240 mm.); tail 5⅛–6¼ inches (130–160 mm.); over-all about 14–16 inches (350–400 mm.); hind foot 1⅞–2⅛ inches (48–55

Fig. 78.—Franklin's ground squirrel.

mm.). Adults in early summer weigh slightly more than a pound (380–490 gm.).

The skull is 52–55 mm. (about 2⅛ inches) long; the rostral portion or snout of the skull is broad and elongate (nasals more than 17 mm. long); the postorbital processes are angled backward, fig. 73b. Dental formula: I 1/1, C 0/0, Pm 2/1, M 3/3.

Life History.—The Franklin's ground squirrel lives in habitats that are similar to those of the thirteen-lined ground squirrel except that generally they contain taller and thicker grass and more brush. It does not often stray as far into the open as does the thirteen-liner; its shyness and its preference for thicker cover make it seem relatively scarce, although it is sometimes seen scurrying across a highway.

Young, usually numbering four or five in a litter, are born about mid-May. They develop rapidly and by July are foraging for themselves. Like their parents, they relish grasses, other herbaceous plants, seeds, and some insects. Members of this species may sometimes be seen feeding on carcasses of animals killed on highways. They are said to kill and feed on other animals, and have been accused of eating eggs occasionally. By fall, they have stored up great quantities of fat in their bodies to tide them over the long winter months.

In its hibernation, this animal is similar to the thirteen-lined ground squirrel. In October or November, it enters hibernation, usually not to return above ground until March or April.

Hawks, badgers, weasels, dogs, man, and other enemies take their toll of the Franklin's ground squirrel; speeding cars on highways may take the greatest toll.

Signs.—Franklin's ground squirrel tracks, sometimes seen along dusty roadsides, are quite similar to those of gray tree squirrels. The print of a hind foot measures about 2 inches and shows five toes, but the print of a front foot, which is shorter, shows only four toes. The location of these tracks, often far from trees, helps to distinguish them from tracks of the gray squirrel.

The burrow of this animal is usually in tall grass or a weedy spot. It is about 3 inches in diameter, and may have a mound of dirt spread fanwise from it. Usually it is better concealed and larger than that of the thirteen-lined ground squirrel.

Distribution.—Franklin's ground squirrel is common in what was originally prairie regions in about the northern two-thirds of Illinois. This ground squirrel is unknown or rare south of an imaginary line connecting Clark and Madison counties. No subspecies has been described. The known total range extends from northwestern Indiana and central Illinois northward to central Manitoba, Saskatchewan, and Alberta, and southward into Kansas and Missouri, fig. 77. It is known in Wisconsin from the southern part of the state.

TAMIAS STRIATUS (Linnaeus)
Eastern Chipmunk

Description.—The eastern chipmunk, fig. 79, is one of the most colorful Illinois mammals. It is russet on the head, flanks, and rump, and grayish on the sides and back; it has five black and two light buff stripes on the back. Stripes are present, but less distinct, on the sides of the face. The under parts are whitish. The tail, which is bushy and about half as long as head plus body, is about the same color on the upper side as the back and is russet on the under side. Large pouches are present within the cheeks.

A characteristic call, consisting of a low-pitched *cluck* or *chuck,* or a shrill chirp, is usually heard before the chipmunk is seen. Frequently this chipmunk twitches its tail in unison with the call.

Length measurements: head and body 6¼–7 inches (160–175 mm.); tail 3¼–4¼ inches (80–110 mm.); over-all about 9½–11¼ inches (240–285 mm.); hind foot 1⅜–1½ inches (34–38 mm.). Weight: about ¼ pound (100 gm.).

The skull, which is 39–42 mm. (about 1⅝ inches) long, lacks an antorbital canal and has only four cheek teeth on each side of the upper jaw. Dental formula: I 1/1, C 0/0, Pm 1/1, M 3/3.

Life History.—The eastern chipmunk is usually unsociable and shy. A chipmunk, at the approach of another of its kind, may give a few flicks of its tail and then dash out suddenly to chase the intruder away. This animal inhabits brushy woods, fig. 1, wooded bluffs, or woods opened by lumbering activities where there is an abundance of old logs and tumbled stones to furnish shelter for burrows, nests, and lookout stations. It will abandon closely pastured woodlands, and its absence from many wooded parts of Illinois is attributed to overgrazing.

The chipmunk is primarily a ground dweller, living in a burrow which it has dug among tree roots or beneath a log, rock, or building. It may occasionally climb into trees. Its burrow may be 20 feet long and have one or several storage chambers in addition to the nest. It feeds chiefly on a variety of fruits and seeds and on insects. In the summertime when juicy berries are ripe, tell-tale stains may be seen on its cheeks. The chipmunk may carry in its cheek pouches a multitude of seeds such as hazelnuts, acorns, basswood fruits, oats, hickory nuts, grass seeds, corn, and elm seeds to store in its chambers for the winter supply of food. It does not grow excessively fat as winter ap-

Fig. 79.—Eastern chipmunk.

proaches, since it will not need to rely on a reserve supply of fat
for nourishment when food becomes scarce but instead will
retire to a well-filled pantry. It becomes dormant during excessively cold weather at any time between November and April
but may come above ground during warm periods. It does not
seem to hibernate in the sense that ground squirrels do.

It is possible for a pair of chipmunks to rear two litters of
young each year. The first litter usually arrives in late April
and the second about August. There are five or six young in
each litter. When a month old, the young are nearly two-thirds
grown and appear outside the burrow.

The chipmunk is preyed upon by hawks, weasels, foxes, cats,
and other animals. Probably its greatest enemy is man, who
through his agricultural endeavors has destroyed much of its
natural habitat.

Signs.—Although the chipmunk sometimes emerges in winter
and sets foot in snow, the best places to look for its tracks
are dusty forest paths. The tracks are somewhat like tracks of
squirrels but smaller and with less tendency for the prints of
the front feet to pair.

The burrow is less than $1\frac{1}{2}$ inches in diameter. The entrance characteristically has no mound when finished and is
usually hidden away under a log, a stone pile, or roots of trees.

Distribution.—The eastern chipmunk may occur in all counties of Illinois but usually it is restricted to unpastured woods
in hilly regions. It is most abundant in rocky, wooded ravines.
Two subspecies are believed to be present in Illinois, *Tamias
striatus ohionensis* Bole & Moulthrop in the Wabash and Ohio
river valleys and *T. s. griseus* Mearns in the remainder of the
state. The range of the species embraces most of southern
Canada from eastern Quebec to southern Manitoba and most
of eastern United States except the far south.

TAMIASCIURUS HUDSONICUS (Erxleben)
Red Squirrel*

Description.—The red squirrel, fig. 80, is smaller than the
fox squirrel or the gray squirrel. It is reddish gray on the back
and whitish on the belly; it has a black line on the sides between

---

*This squirrel should not be confused with the fox squirrel, which is sometimes
called the red squirrel in Illinois.

the upper and under parts. The bushy tail is not quite so long as the body and is reddish gray above, like the back, and yellowish gray below. In the winter pelage, ear tufts are apparent.

Length measurements: head and body 7¼–8¼ inches (186–210 mm.); tail 4⅛–5⅛ inches (104–130 mm.); over-all 11½–13½ inches (290–340 mm.); hind foot 1¾–2⅛ inches (44–53 mm.). Weight: approximately ⅖ pound (140–220 gm.).

The skull is 43–49 mm. (about 1¾ inches) long. The third upper premolar, if present, is very small, and the upper cheek teeth are less than 10 mm. long. Dental formula: I 1/1, C 0/0, Pm 2/1 or 1/1, M 3/3.

Fig. 80.—Red squirrel.

Life History.—In former years, the red squirrel probably frequented the hardwood and coniferous forests of northern Illinois.

This squirrel does not hibernate, although it becomes inactive during periods of extreme cold. It sometimes makes long burrows in snow, probably in search of food. Usually it makes its nest of grasses and leaves in a cavity of a tree, but sometimes it makes a leaf nest among the branches.

Distribution.—Although it almost certainly no longer inhabits Illinois, the red squirrel formerly occurred in scattered colonies in the northern part of the state. There are four authentic records of this squirrel for the state: Lake Forest and Fox Lake in Lake County, Lawnridge in Marshall County, and Hennepin in Putnam County. The specimens on which these records are based (all collected before 1912) belong to the subspecies *Tamiasciurus hudsonicus loquax* (Bangs). The species occurs throughout most of Canada and Alaska; in western United States in mountainous areas as far south as southeastern Arizona; in eastern United States as far south as southern Iowa, central Indiana, and western North Carolina.

SCIURUS CAROLINENSIS Gmelin
Eastern Gray Squirrel

Description.—The eastern gray squirrel, fig. 81, generally is grayish on the back and sides and has a wash or overlay of fulvous or yellow on the sides and legs. The belly is white or light gray, and the bushy tail is gray, tipped with white. Tufts of white hair behind the ears and a ring of white around each eye are characteristic.

Melanistic ("black") or albinistic ("white") individuals or colonies of this species are occasionally seen. In former years, melanistic individuals were frequently found, especially in the northern part of Illinois. In the middle 1800's, an entire group, numbering nearly 50 individuals, near the Rock River consisted of "blacks." At present, there is a sizable colony of albinistic individuals established in and near Olney.

Length measurements: head and body 8¼–10½ inches (210–270 mm.); tail 7½–9½ inches (190–240 mm.); over-all 16–20 inches (400–510 mm.), males averaging about 18¼ inches (464 mm.), females 17¼ inches (439 mm.); hind foot 2⅜–3 inches (60–75 mm.). Average weight: about 1¼ pounds.

The skull is 59–64 mm. (about 2⅜ inches) long. The third upper premolar on each side is small and peglike. Dental formula: I 1/1, C 0/0, Pm 2/1, M 3/3.

Life History.—The gray squirrel is commonly seen in trees bordering city streets, in parks, on lawns, near wooded streams and heavy stands of timber, and in areas where there is abundant ground cover and brush not greatly disturbed by ax or livestock. Probably it spends more time in trees and less time on the ground than does the fox squirrel; it is more adept and graceful in moving through trees.

The gray squirrel lives in a leaf nest or in a hollow within the main trunk of a tree. A litter of three to five young is brought forth in the nest about the middle of February in south-

Fig. 81.—Eastern gray squirrel.

ern Illinois and about the middle of March in northern Illinois. A second breeding period begins about June 15 in southern Illinois, about June 25 in central Illinois, and about July 5 in northern Illinois. About 45 days later, the second litter is born.

Food of the gray squirrel consists chiefly of buds, seeds, acorns, nuts, and other fruits of nearly all trees in the habitat where the animal lives. It consists also of fungi, corn kernels, soybeans, berries, and grapes. This squirrel stores quantities of food for winter use, as it does not hibernate. It may, however, become inactive and remain within its nest during severe cold spells.

Early in the settlement of Illinois, grays were apparently much more abundant than they are today. In some years, the production of nuts and seeds was especially good, and ample food resulted in a large number of squirrels. In those years in which the nut crop was a failure, grays in large numbers would move as a group across country, surmounting many barriers en route. This forced migration resulted in their being called "migratory squirrels" by some people.

The gray squirrel and the fox squirrel usually do not live together in Illinois. In some towns there are only grays and in other towns only fox squirrels. In some timber, usually open timber, there are only fox squirrels; in other timber, usually that with ample undercover, there are only grays. These two species do not cross or interbreed in nature, despite stories to the contrary.

Signs.—Prints of the hind feet of the gray squirrel measure a little less than 2 inches each over-all, and prints of the front feet are shorter. Tracks of the gray squirrel resemble those of the fox squirrel, fig. 34. Gnawed remains of nuts and depressions an inch or so deep and about as wide among fallen leaves in brushy woodlands are signs of the gray squirrel.

Leaf nests, situated well up in trees, in a type of habitat described above, indicate the presence of gray squirrels; a single squirrel may be the owner of more than one leaf nest.

Distribution.—The gray squirrel is fairly common in wooded areas of Illinois. Individuals in the northern two-thirds of the state are of the subspecies *Sciurus carolinensis pennsylvanicus* Ord, those in the southern third *S. c. carolinensis* Gmelin. The species occurs in most of the eastern half of the United States and in southern Ontario.

SCIURUS NIGER Linnaeus
Eastern Fox Squirrel

Description.—The eastern fox squirrel, fig. 82, generally is rusty-yellow, with a mixture of gray, on the upper parts and reddish yellow on the under parts. The cheeks, the fur behind the ears, and the feet are a light reddish yellow. The bushy tail is bordered with reddish.

Length measurements: head and body 11–11¾ inches (280–300 mm.); tail 8¾–10¼ inches (220–260 mm.); over-all 19½–22 inches (500–560 mm.); hind foot 2⅜–3 inches (60–78 mm.). Weight: average approximately 1¾ pounds, about half a pound more than that of the gray squirrel.

The skull is 60–70 mm. (about 2½ inches) long; there are only four cheek teeth on each side of the upper jaw (usually five in gray squirrel). The zygomatic arches are nearly parallel, fig. 74k. Dental formula: I 1/1, C 0/0, Pm 1/1, M 3/3.

Fig. 82.—Eastern fox squirrel.

Life History.—The fox squirrel is familiar to most people in Illinois, for it occurs in many of the city parks and farm wood lots of the state. Since it prefers woods with openings, the removal of timber for lumber or the grazing of woodlands by cattle has not adversely affected the fox squirrel so much as the gray squirrel. Wood lots of even a few acres provide suitable habitat if they are connected by hedgerows or scattered trees, and if they have trees with cavities and trees that produce nuts or other seeds.

The fox squirrel lives in a leaf nest, fig. 9, or in the cavity of a hollow tree. It has two periods of mating and breeding, one in midwinter (December or January) and another in early summer (May in southern, June in northern Illinois). Young numbering two to five per litter are born in February or March and in July or August. A female of 2 years or older may have two litters a year, but a yearling has only one litter.

The fox squirrel feeds on nuts, fruits of the oak, elm, and beech, corn, tree buds, mushrooms, and even birds' eggs. Mast is important in fall and winter; fleshy fruits, buds, and left-over mast in spring; many items in summer. Nuts for winter use are individually buried in small pits the animal digs in the ground. The fox squirrel does not hibernate; it hunts out, actually smells out, these food reserves when winter comes. Because of its habit of burying food, the fox squirrel spends much time on the ground and frequently moves considerable distances from trees.

Signs.—Fox squirrel tracks, fig. 34, average slightly larger than gray squirrel tracks, and the small prints of the front feet are more often paired. Their location helps to distinguish tracks of the two species. If the tracks are along hedgerows, along narrow strips of timber, or among widely scattered trees in pastures or cultivated fields, they generally belong to the fox squirrel. If they are in very dense stands of brushy timber, they are likely to belong to the gray squirrel. The leaf nest of the fox squirrel resembles that of the gray squirrel.

Distribution.—The fox squirrel occurs throughout Illinois. Individuals in this state are of the subspecies *Sciurus niger rufiventer* Geoffroy. The range of the species extends from Delaware and southern Pennsylvania westward through southern Minnesota, Nebraska, and much of Texas, and southward to the Gulf of Mexico.

Fig. 83.—Southern flying squirrel.

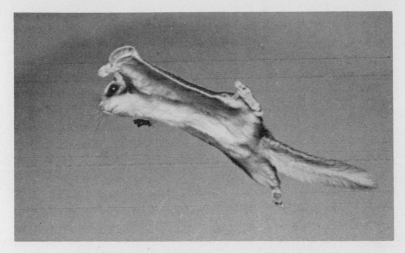

Fig. 84.—Southern flying squirrel in "flight."

GLAUCOMYS VOLANS (Linnaeus)
Southern Flying Squirrel

Description.—The southern flying squirrel, fig. 83, is about the size of the eastern chipmunk but appears larger than it really is because of its "flying" membranes, thick fur, and broad tail. Its eyes are especially large and its ears project only a short distance above the fur. The "flying" membrane on each side of the body consists of a loose fold of skin extending from wrist to ankle. The glossy fur on the back and sides of the body is gray, drab, or pinkish cinnamon; the under parts are pure white. The feet are a dusky color. The flattened tail is about the same color as the back.

Length measurements: head and body 5⅛–5½ inches (130–140 mm.); tail 3½–4¼ inches (90–110 mm.); over-all 8⅝–9¾ inches (220–250 mm.); hind foot about 1⅛ inches (27–33 mm.). Weight about 2 ounces (40–70 gm.).

The skull is 33–36 mm. (about 1⅜ inches) long. The interorbital region is deeply notched, fig. 74*j,* and the postorbital processes are large. Dental formula: I 1/1, C 0/0, Pm 2/1, M 3/3.

Life History.—The flying squirrel is abroad only at night and sleeps by day in a cavity within a tree. It is a very intriguing animal, for it has perfected the art of gliding (loosely

referred to as flying). To get from one tree to another, perhaps 30 or more yards away, a flying squirrel does not need to
go to the ground and expose itself to dangers there, fig. 1. Instead, it leaps, spreads its legs so that the fold of skin on each
side of the body is fully extended, fig. 84, and, aided by the flattened tail, glides to the trunk of a distant tree. The direction
of the glide can be controlled somewhat by movement of the
tail and the membrane. The squirrel does not fly in the sense
that it flaps the membrane, nor is its point of landing ever higher
than its point of departure.

The southern flying squirrel is found in areas that are heavily
wooded, but not so heavily wooded as to prevent gliding, and
that have trees with ample woodpecker holes or similar cavities
in which to make nests. In one of these cavities a female may
have one or two litters of two to six young each year. The first
litter is born in late March or early April and the second in
August. The flying squirrel feeds on nuts, seeds, tree buds,
fruits, insects, and birds' eggs.

This is the commonest squirrel in much of the heavy timber
of Illinois. Noisy taps on a hollow tree or on a tree with woodpecker holes may cause a sleepy flying squirrel to poke its head
out of one of the cavities. Several individuals may occupy one
cavity, for these animals are quite sociable. They make ideal
pets.

Signs.—Tracks of the flying squirrel are rarely seen. The
footprints are like those of other tree squirrels. Prints of the
hind feet are each about 1¼ inches long. Opened nuts beneath
an old woodpecker hole in a tree may indicate the presence of
one or more flying squirrels. Nuts opened by a flying squirrel
have a roughly circular opening and are not largely gnawed
away, as are those opened by the gray squirrel or the fox squirrel. Tooth marks of the flying squirrel are finer than those of
the other tree squirrels common in Illinois.

Distribution.—The southern flying squirrel occurs in mature
woodlands of Illinois, and probably it is common in such habitats. Individuals in this state are of the subspecies *Glaucomys
volans volans* (Linnaeus). The species has a disjunct range;
one population occupies approximately the eastern half of the
United States (except northern New England); one occurs in
the central part of the Central American countries; a third
occurs in the central part of the Mexican Plateau.

GEOMYS BURSARIUS (Shaw)
Plains Pocket Gopher*

Description.—The plains pocket gopher, fig. 85, in Illinois usually is slate gray to black on the back, light gray on the under parts, and white on the nose, feet, and terminal half of the tail. In other parts of its range, this gopher is principally brown or tawny, but in Illinois only rarely is one found that is chocolate brown on the back. The eyes are small, and the ears are round and extremely short. The front legs are stout, and the claws of the front feet are long. The tail is about one-third the length

Fig. 85.—Plains pocket gopher.

of the head plus body and sparsely covered with short hairs. An external pocket or pouch is present on each side of the mouth. The lips can be closed behind the long, heavy upper incisors.

Length measurements: head and body of males 7¾–8¾ inches (200–222 mm.), of females 7–8¼ inches (180–210 mm.); tail 2½–4 inches (65–100 mm.); over-all for males 10½–12¾ inches (265–322 mm.), for females 9½–12¼ inches (245–311 mm.); hind foot about 1⅜ inches (30–37 mm.). Weight: about ¾ pound (300–400 gm.).

The skull is strongly ridged for muscle attachments and is 49–59 mm. (about 2⅛ inches) long in males, 46–56 mm. (about 2 inches) in females. The rostrum is heavy and broad. The incisors are strong, and each upper incisor has at least two grooves on its front surface, fig. 73c. Dental formula: I 1/1, C 0/0, Pm 1/1, M 3/3.

*Not to be confused with the ground squirrels, sometimes called gophers.

Life History.—The plains pocket gopher lives continuously, or nearly so, below ground and comes to the surface only to dump earth from its burrow or to make very short forays in quest of food. Normally, it breeds, nests, feeds, and carries on all other activities below ground. Its burrow system is several hundred feet long, and each animal throws up many mounds or piles of dirt as it digs its burrow system.

Except for a short time during the breeding season, each burrow system is occupied by a single individual. The tunnels usually are not deeper than 4 feet; they have a main nest, and possibly subsidiary nests, and several storage chambers of food. In the storage chambers, the gopher stores roots, stems, and leaves of sweet clover, alfalfa, dandelion, plantain, and other herbs, as well as roots of shrubby plants. It digs innumerable side tunnels to the roots of likely food plants. The gopher cuts the roots in appropriate lengths, stuffs the pieces adeptly into its cheek pouches, and carries them to its storage chamber.

This gopher has certain adaptations for a lifetime within a narrow, dark tunnel. The tip of the tail, which is highly sensitive, is employed as a guide when the animal is backing. With the tail projecting straight back, except for a little crook at the end which causes the tip to touch the wall or floor of the tunnel, the gopher can run backward almost as swiftly and surely as it can run forward. The skin of the gopher is loose and easily stretched, allowing the animal to reverse itself readily within a narrow tunnel. The lips are modified in such a way that they can be closed behind the incisors and thus keep dirt out of the mouth while the animal is gouging out soil or cutting roots. The long, curved claws aid in digging and pushing dirt from the tunnel.

The pocket gopher in Illinois probably has only one litter a year, born some time between early March and early May. There may be as many as six young in a litter, but the usual number is four. The mother, having driven the male from her burrow system soon after the breeding season is over, raises the young by herself.

Weasels, badgers, and snakes, particularly bull snakes, are the principal enemies of the pocket gopher. The tunnel entrances are kept solidly blocked with dirt at almost all times and this may discourage enemies from entering.

Signs.—Mounds of dirt in grasslands, in hayfields, or along grassy roadsides in central Illinois may be signs of the pocket gopher. At a distance, gopher mounds may be mistaken for those of the mole. The gopher produces a mound, figs. 2 and 7, by pushing the dirt from the mouth of the sloping entranceway in such a manner that this dirt forms a relatively low, fan-shaped mound. The mouth of the entranceway, which is on the low side of the mound, is plugged by the last "load" of dirt; the "plug" is a distinguishing characteristic of a recently constructed pocket gopher mound. The mole produces a mound, fig. 6, by pushing the dirt straight up from beneath the surface of the ground; the loosened dirt cascades down evenly to form a cone with relatively steep sides, like those of a volcano. In weathered mounds, these differences are difficult to detect. A gopher mound is 12 or more inches across, and a mole mound is about 10 inches. The pocket gopher never has subsurface runs that hump up the dirt radiating out from the mound as does the mole.

Distribution.—The pocket gopher is abundant in the sandy and black soils east and south of the Illinois and Kankakee rivers, and in Madison County, fig. 86. In Illinois is the subspecies *Geomys bursarius illinoensis* Komarek & Spencer. The range of the species includes an irregular area in the center of the North American continent, fig. 86.

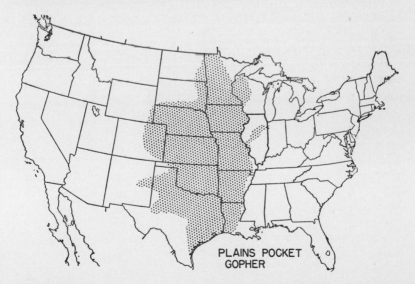

Fig. 86.—Known distribution, in the United States, of plains pocket gopher.

CASTOR CANADENSIS Kuhl
Beaver

Description.—The beaver, fig. 87, is a large, dark brown, semiaquatic mammal with a flat, paddle-shaped, scaly tail, webbed hind feet, and small ears.

Length measurements: head and body 26–31 inches (660–790 mm.); tail 9½–10¼ inches (240–260 mm.); over-all about 35–41 inches (900–1,050 mm.); hind foot 6½–7 inches (165–180 mm.). Weight: usually 25–50 pounds; a weight of 78¼ pounds is recorded for a specimen taken on the Mississippi River near Meyer, Adams County.

The skull is massive and about 120 mm. (4¾ inches) long. The large, red-orange incisors are well suited for gnawing, and the back cheek teeth well suited for grinding, fig. 73d. Dental formula: I 1/1, C 0/0, Pm 1/1, M 3/3.

Life History.—The beaver needs a continuous supply of water and a supply of suitable food near water. For food it prefers poplar, maple, birch, and willow; it may eat other kinds of trees, as well as cattails and other aquatic plants. The trees and branches cut by a colony of beavers and transported to their dam serve not only as a means of retaining the water but also as a pantry for food.

A colony of beavers consists usually of an adult male and female and their offspring of the year and the previous year. Since

Fig. 87.—Beaver.

the usual number in each litter is four, there may be as many as
10 beavers in a colony. As the older litter approaches the age of
2 years, it is driven from the colony. This forced dispersion
results in the continuous establishment of new colonies, some
nearby, others far removed. During this dispersion, adult
beavers may be killed by dogs, men, or cars. Young beavers
may be killed by minks or by extreme fluctuations in water level.
By and large, beavers have few natural enemies in Illinois.

Some beavers live in burrows in banks of lakes or streams.
Many others live in lodges behind their dams. Their dams may
be as much as 6 feet in height and 200 feet in length and so constructed as to withstand heavy floodwaters. They are made of
tree branches, sticks, mud, and stones, and are plastered with
mud on the upstream side. Their great lodges, fig. 11, built of
similar material, often extend 6 feet above water and are several
times as wide as high. From dusk to daylight, beavers work on
their dams or lodges—adding branches, replacing mud, or felling
nearby trees. Each member of a colony except the young of the
year is "busy as a beaver."

Signs.—Some of the most dramatic wildlife signs in Illinois
are those of the beaver. In addition to their dams and lodges,
beavers leave such signs as felled trees, sometimes more than a
foot in diameter, gnawed down for food, or the stumps of trees
showing the characteristic wide tooth marks, fig. 28*e*; great piles
of branches stored near the lodges for food; canals several feet
deep and several feet wide down which they float timber; large
burrows 12 to 18 inches in diameter dug in stream banks; and
footprints, chips, droppings, nibbled cornstalks, and a host of
lesser signs. Branches cut into sections several feet long with
bark peeled from them may indicate the presence of beavers.

Beaver droppings are loosely packed, 1–1½ inches in length,
and composed principally of wood fiber. They are light and float
on the surface of the water only a brief time before disinte-
grating. They are not deposited on stones or logs in or near
water, as are the droppings of muskrats.

The print of a hind foot of the beaver, fig. 28*b*, measures 6
inches in length and 5 inches across, and it shows the web connecting the toes. The print of a front foot, fig. 28*a*, roughly
half as long as that of a hind foot, shows long claw marks.

Distribution.—In former times, the native beaver was abundant in Illinois. Probably *Castor canadensis michiganensis*

Bailey occurred over most of the state, and *C. c. carolinensis*
Rhoads inhabited the extreme southern part. By the middle
1800's beavers had been reduced to the point of being rare in the
state, and during the late 1800's or early 1900's the native population was exterminated. In 1935, beavers brought from Wisconsin were released in Illinois; subsequently there were inva-
sions of beavers from Iowa and Indiana. These animals increased in numbers and extended their range until by 1954
beavers were present in nearly half the counties of Illinois;
through transplantations or natural movements there were
nearly 600 colonies. All members of the species occurring in
Illinois at present probably belong to the subspecies *C. c. michiganensis*. The natural range of the species occupies most of
Alaska, Canada, and the United States except parts of the
southeastern and Gulf Coast states and the arid parts of the
southwestern states.

REITHRODONTOMYS MEGALOTIS (Baird)

Western Harvest Mouse

Description.—The western harvest mouse, fig. 4, smaller
than the white-footed mouse, is similar to the house mouse except
in color. Its back and sides are brown, and its under parts are
silvery gray; the under side of its tail is whitish, and the upper
sides of its hind feet are white. Each upper front tooth (incisor)
has a groove down its face. The only other rodents now reported
in Illinois with such grooves are the plains pocket gopher, the
meadow jumping mouse, and the bog lemming.

Length measurements: head and body 2½–3 inches (66–77
mm.); tail 2¼–2⅝ inches (57–68 mm.); over-all 4¾–5¾
inches (123–145 mm.); hind foot about ⅝ inch (16–18 mm.);
ear from notch about ½ inch (12–14 mm.).

The skull averages 20.3 mm. (about ¾ inch) in length; in
width across the zygomatic arches it averages 10.4 mm. (about
⅜ inch). Dental formula: I 1/1, C 0/0, Pm 0/0, M 3/3.

Life History.—Western harvest mice taken in Illinois
near Mount Carroll were found living among brome-grass,
goldenrod, blackberry, ragweed, and bluegrass. The vegetation
was tall but not thick. Harvest mice were quite rare in this
habitat and were associated with masked shrews, short-tailed
shrews, white-footed mice, and bog lemmings.

Little is known about the life history of the harvest mouse in Illinois. In some other states, this species is known to build a globular nest in dense grasses or weeds, rather than under logs or below the surface of the ground. It feeds on seeds, is a good climber, and frequently crawls far up into plants in search of food. Breeding begins in the spring. Young may appear by late April and when 3 weeks old are able to feed by themselves.

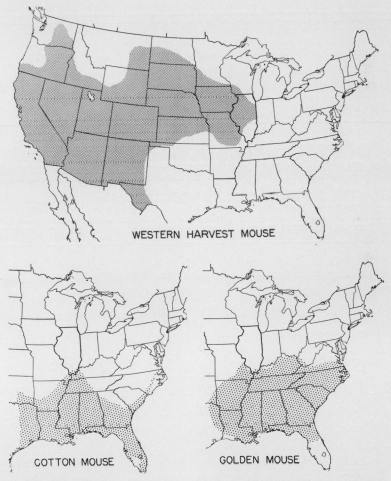

WESTERN HARVEST MOUSE

COTTON MOUSE

GOLDEN MOUSE

Fig. 88.—Known distribution, in the United States, of some mice with a limited range in Illinois.

Signs.—Tracks or droppings of the western harvest mouse cannot easily be differentiated from those of other mice. A grassy nest above the ground in a weedy or grassy field may be that of a harvest mouse.

Distribution.—In Illinois the western harvest mouse has been taken in Carroll County. Elsewhere east of the Mississippi River, it has been recorded from southern Wisconsin. It is found in much of the United States west of the Mississippi and in Mexico, fig. 88. Animals of this species in Illinois have been referred to the subspecies *Reithrodontomys megalotis dychei* Allen.

REITHRODONTOMYS HUMILIS (Audubon & Bachman)
Eastern Harvest Mouse

Although specimens or records of this species are lacking for Illinois, an eastern harvest mouse was caught recently across the Ohio River near Paducah, Kentucky. Because of this proximity, the species might be expected in brome-grass and savanna-like habitats in extreme southeastern Illinois.

The combination of grooved incisors and moderately long tail will serve to identify the genus. Differences in their geographic ranges can be used as a provisional means of distinction between the eastern harvest mouse and the paler but similar western harvest mouse, *Reithrodontomys megalotis*.

PEROMYSCUS MANICULATUS (Wagner)*
Deer Mouse   Prairie Deer Mouse

Description.—The smallest Illinois *Peromyscus,* the deer mouse, fig. 89, usually is grayish brown (rarely ochraceous) on the upper parts, with a blackish area, more or less concentrated in a band, down the middle of the back. The tail, except for a narrow stripe of grayish brown on top, the feet, and the under parts are whitish. The ears are grayish brown, margined with cream or white. The hind feet are unusually small, and the tail is relatively short.

Length measurements: head and body 3⅛–3½ inches (81–90 mm.); tail 1¾–2⅜ inches (44–60 mm.); over-all 5–6 inches

---

*As used here, the name white-footed mouse is restricted to the species *Peromyscus leucopus.*

(125–150 mm.); hind foot ⅝–¾ inch (15–18 mm.); ear from notch about ½ inch (11–14 mm.).

The skull is 22.0–24.5 mm. (about ⅞–1 inch) long; its width across the zygomatic arches just behind the infraorbital plate is less than 12 mm. Dental formula: I 1/1, C 0/0, Pm 0/0, M 3/3.

The deer mouse may usually be distinguished from the more abundant white-footed mouse on the basis of the following combination of characters: hind foot less than 19 mm. long, rather than 19 mm. or more as in the white-footed mouse; tail less than 60 mm. long, rather than more than 60 mm.; total length

Fig. 89.—Deer mouse.

of adults less than 155 mm., rather than more; width of skull across zygomatic arches just behind the infraorbital plate less than 12 mm., rather than more; length of skull 24.5 mm. or less, rather than more.

Life History.—The deer mouse is an inhabitant of the prairie and is most abundant in areas of ungrazed and uncut grass and forbs, such as grow along railroad and highway rights-of-way, in extensive weedy fields, and along fencerows, fig. 2. It is at home under a haystack or a corn shock, among dry weeds along a fence, or under a rock or board.

The nest, usually underground, is approximately the size of a man's two cupped hands. It is formed of a coarse outer frame-

Fig. 90.—White-footed mouse.

work of stems, roots, and leaves, and holds a soft inner lining
of plant down, fur, or feathers.

Young may be produced in nearly any month of the year. An
adult female may have several litters (potentially a dozen, but
probably only four or five) each year, and there are usually four
young in each litter. The young, blind and naked at birth, grow
rapidly. Females can start breeding when only 5 to 10 weeks
old.

The deer mouse feeds on seeds of grasses and weeds, and on
berries, buds, insects, and possibly some green plants. It does
not hibernate. Probably it hoards some seeds in burrows or
tunnels near its nest for winter use. This mouse is preyed upon
by owls, snakes, weasels, foxes, and nearly all other Illinois fur
bearers. In the wild, it probably never lives for more than 2
years before it meets an untimely end; it may live as long as 9
years in captivity.

Signs.—It is difficult to tell the footprints and droppings of
the various kinds of *Peromyscus* apart. The location of signs
may aid in determining the kind of mouse that made the signs,
as habitats of the various species differ considerably.

Prints of the hind feet of the deer mouse are paired and those
of the front feet are nearly so. The sets, fig. 31, are about 3
inches apart when made by a mouse bounding at an ordinary
gait; much farther apart when made by a speeding mouse. Droppings are brown or black and between ¼ and ⅜ inch long.

Distribution.—The deer mouse may be locally abundant,
particularly in sand prairies. Colonies probably occur throughout the state. The subspecies in Illinois is *Peromyscus maniculatus bairdii* (Hoy & Kennicott). The main range of the species extends from Labrador to central Alaska and southward to
southern Mexico in the west and Tennessee in the east; the
range also extends down the Appalachians to northeastern
Georgia.

PEROMYSCUS LEUCOPUS (Rafinesque)

White-Footed Mouse   Woodland White-Footed Mouse

Description.—The white-footed mouse, fig. 90, has the same
proportions as the house mouse, but is larger. In the adult, the
upper parts are bright brown or fulvous, the under parts white.
The ears are dusky brown, with whitish edges. All four feet

are white. The tail is bicolored, dark on the upper part, whitish on the under part. In the immature mouse, the upper parts are a dull brown or gray, the under parts white.

Length measurements: head and body 3–4 inches (78–100 mm.); tail 2⅜–3⅜ inches (60–85 mm.); over-all 5⅜–7¼ inches (138–185 mm.); hind foot ¾–⅞ inch (18–22 mm.); ear from notch ⅝–¾ inch (15–19 mm.).

The skull is 24.5–28.0 mm. (about 1–1⅛ inches) long; the width across the zygomatic arches just behind the infraorbital plate is more than 12 mm. Part of the skull is shown in fig. 74h. Dental formula: I 1/1, C 0/0, Pm 0/0, M 3/3.

The white-footed mouse closely resembles the deer mouse and the cotton mouse.

Life History.—The white-footed mouse lives in forests, fig. 1, brushlands, river bottoms, forest edges, and even in brushy areas extending out into prairies. Probably 3 to 12 mice, sometimes even more, per acre occur in these habitats. A white-footed mouse may be found under a log, within a stump, in a once-abandoned bird's nest, or in a shallow burrow. It may make its nest in, on, or above the ground. If a home in a burrow proves too damp, the white-footed mouse may move into a decaying log or into a tree stump. In the fall, it may move to an unused bird's nest, such as that of the goldfinch, after working diligently a few nights to build a dome over it and arrange a soft lining. It may nest in a woodpecker hole or a bird box near timber; it rarely enters inhabited buildings.

Breeding may take place in all except the very coldest months of the year, but the majority of young are produced in spring, early summer, and fall. The white-footed mouse is probably as prolific as the deer mouse. A mature female gives birth yearly to at least four litters, with four or five young in each litter. The young are hairless and blind at birth. They grow rapidly and become independent of family ties probably when they are scarcely more than a month old.

The white-footed mouse is nearly omnivorous. It normally feeds on seeds of wild herbaceous plants, nuts, buds, fruits, and insects; if given an opportunity it eats grains and pantry items.

Signs.—Tracks and droppings of the white-footed mouse resemble those of the deer mouse, fig. 31, but they are slightly larger. The home of the white-footed mouse is described above under life history.

Distribution.—The white-footed mouse is abundant throughout Illinois. The subspecies in northern and central Illinois is
Peromyscus leucopus noveboracensis (Fischer); that in southern
Illinois is probably *P. l. leucopus* (Rafinesque). The range of
the species extends from southern Maine to southern Alberta
and southward to southern Mexico. Its western limits are
marked by northeastern Wyoming, northwestern Kansas, and
south-central Arizona. The range does not include the extreme
southeastern United States.

PEROMYSCUS GOSSYPINUS (Le Conte)
Cotton Mouse

Description.—The adult of the cotton mouse is dark reddish
brown on the upper parts and white on the under parts. An
immature may be gray or grayish brown on the upper parts.
The adult closely resembles the white-footed mouse but usually
can be distinguished from it by the longer body and hind feet and
the larger skull.

Length measurements: head and body 4–4¼ inches (100–107
mm.); tail 3–3½ inches (78–88 mm.); over-all 7–7⅝ inches
(178–195 mm.); hind foot about 1 inch (23–26 mm.); ear from
notch ⅝ inch (15–16 mm.) in dry study skins.

The skull is 28–30 mm. (about 1⅛ inches) long. The upper
molar tooth row is 3.6–4.0 mm. (about ⅛ inch) long. Dental
formula: I 1/1, C 0/0, Pm 0/0, M 3/3.

Life History.—The cotton mouse lives in swamps and bottomlands, and in forests adjacent to them, fig. 3. Apparently
it picks the driest spot beneath a fallen log or stump to place
its nest.

Little is known about the cotton mouse in Illinois. Probably
its breeding habits are similar to those of the white-footed
mouse. Both species may occupy the same swampy woodlands.

It is not clear how the common name for this mouse was
derived; in southern states where it is abundant it usually is
found in areas not far removed from cotton fields.

Signs.—Tracks and nests of the cotton mouse are like those
of the white-footed mouse.

Distribution.—The cotton mouse is uncommon in Illinois and
is known in the state only in the southern tip south of the Ozark
Plateau and the Shawnee Hills. The subspecies in this state is

Peromyscus gossypinus megacephalus (Rhoads). The range of
the species is an irregular area that embraces most of the southeastern states, fig. 88. It includes northward extensions into
northeastern Virginia and southern Illinois. Its westward limit
is eastern Texas.

PEROMYSCUS NUTTALLI (Harlan)
Golden Mouse

Description.—The golden mouse, fig. 91, is reddish brown
or golden on the upper parts and white or cream colored on the
lower parts. The ears are red and the feet are white. The
eyes are large and conspicuous, the cheek pouches are thin and
inconspicuous, and the tail is slightly shorter than the body.

Length measurements: head and body $3\frac{1}{4}$–$3\frac{1}{2}$ inches (83–89
mm.); tail $2\frac{5}{8}$–$3\frac{1}{4}$ inches (67–83 mm.); over-all $5\frac{7}{8}$–$6\frac{3}{4}$
inches (150–172 mm.); hind foot $\frac{3}{4}$ inch (18–20 mm.); ear
from notch about $\frac{3}{4}$ inch (16–18 mm.).

The skull is 25–27 mm. (about 1 inch) long. The infraorbital
plate is straight along its front margin, fig. 74*g*. Dental formula: I 1/1, C 0/0, Pm 0/0, M 3/3.

The following combination of characters serves to distinguish
the golden mouse from all other species of *Peromyscus* in Illinois: reddish brown or golden color of both young and adults,
red ears, thin cheek pouches, and straight infraorbital plate.

Life History.—The bright-colored golden mouse is as much
at home in trees, vines, or bushes as it is on the ground. It con-
structs its home in a thicket of honeysuckle, greenbrier, or poison
ivy, or in the crotch or branches of a tree or bush, fig. 3. In
Illinois, it apparently prefers the thick timber bordering cypress
swamps.

Its nest, fig. 91, is about 8 inches in diameter and globular in
shape; the single entrance is closed except when the mouse is
entering or leaving the nest. The inner lining is of soft, finely
shredded material. Several golden mice, probably a family
group, may occupy a single nest at the same time. Young may
be brought forth in a nest at any time between March and
October. A female may have more than one litter each year,
with two or three young in each litter.

The golden mouse spends much of its time in vines and trees,
and it is, as might be expected, an adept, sure-footed climber.

Fig. 91.—Golden mouse and its nest.

Its tail, used as a fifth appendage, is frequently wrapped around twigs to aid the animal in balancing and maneuvering. The golden mouse is a seed eater; it feasts on the soft inner parts of the seeds of sumac, greenbrier, wild cherry, dogwood, pokeweed, clover, bittersweet, and oak. The mouse fills its small cheek pouches with these seeds and carries them to a nestlike feeding platform situated above the ground. Here it may eat the seeds immediately or store them for future use.

The beautifully colored golden mouse makes a docile pet. A few days of confinement and gentle handling will usually result in its complete adjustment and conditioning to man.

Signs.—A globular nest or feeding platform in a thicket or tree may belong to a mouse of this species. However, in many parts of Illinois, such a nest may belong to a white-footed mouse.

Distribution.—The golden mouse, evidently uncommon in Illinois, has been taken only in Alexander and Johnson counties. The Illinois subspecies is *Peromyscus nuttalli aureolus* (Audubon & Bachman). The range of the species extends from southern Virginia west to southern Missouri and eastern Oklahoma and southward to the Gulf Coast and to central Florida, fig. 88.

ORYZOMYS PALUSTRIS (Harlan)
Rice Rat

Description.—The rice rat, fig. 92, is a third to half the size of the Norway rat and has a tail slightly shorter than the body. It is grayish brown on the upper parts and silvery gray on the

Fig. 92.—Rice rat.

under parts. The upper side of the tail is the same color as the back of the animal and the lower side is only slightly lighter. The fur of the under parts of the animal is soft and woolly. In general, the rice rat closely resembles the Norway rat, but the light color of the under parts, including the under side of the tail, is distinctive.

Length measurements: head and body 4¾–5½ inches (120–140 mm.); tail 4¼–5½ inches (110–140 mm.); over-all 9–11 inches (230–280 mm.); hind foot 1¼ inches (30–33 mm.).

The skull of the rice rat is 30–33 mm. (about 1¼ inches) long and has supraorbital ridges along the lateral margins of the braincase. In this rat, the cusps on the upper cheek teeth are arranged in two longitudinal rows, whereas in the Norway rat they are arranged in three rows, fig. 74*a, c*. Dental formula: I 1/1, C 0/0, Pm 0/0, M 3/3.

Life History.—The rice rat is nearly as much at home in the water as the muskrat and almost as good a trail-maker as the meadow vole; its runways thread the dense vegetation of marshes and swamp margins, fig. 3. The rice rat readily takes to water when disturbed, and in many cases it is necessary for it to go through water to reach its nest.

The nest of the rice rat is a globular structure of dry grasses and leaves. It may be constructed in an oval chamber at the end of a burrow, in one of the highest, driest objects in a swamp or bottomland, such as a stump or hollow log or a pile of debris, or on the top of a fence post which is well covered with vines; or it may be suspended on a bunch of interlaced marsh grass or embedded in a tangled mass of blue flags or marsh grasses.

Breeding may extend from March to October. Three to eight young are produced in a single litter. The young are helpless at birth but grow and mature rapidly. Females may begin bearing young when they are about 2 months old.

The rice rat is active during daylight hours as well as at night. In this regard, it resembles the meadow vole. It feeds principally on green plants, but also on young turtles, snails, and crayfish. Apparently this rat acquired its currently used common name because, in parts of its range, individuals of this species were found invading rice fields to feed on newly planted and also maturing grain. Each rat normally eats approximately 25 per cent of its own weight in food every day. The rice rat is fed upon by swampland predators, including snakes, owls, hawks, minks, and raccoons.

Signs.—The runways of the rice rat resemble those of the meadow vole but they are more open and without the numerous cut sections of vegetation lying in them. Feeding platforms of freshly cut grass stems near such runways are almost certain signs of the rice rat. These may be as large as dinner plates, but smaller than those of the muskrat. The burrow of the rice rat is seldom more than a foot deep, and the entrance is commonly a few inches above the high water level usual for the area in which it is located.

Footprints in soft mud near water may be those of a rice rat, especially if they lead to a nest or a feeding platform. The

print of a hind foot is more than 28 mm. (about 1⅛ inches) long
and may be confused with that of the Norway rat.

Distribution.—The rice rat is little known in Illinois, having
been collected only in some of the southernmost counties. The
subspecies in Illinois is *Oryzomys palustris palustris* (Harlan).
The range of the species includes an area that extends from
southern New Jersey westward to northwestern Arkansas and

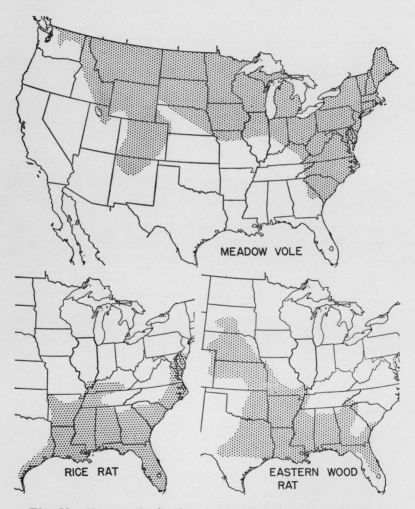

Fig. 93.—Known distribution, in the United States, of some rodents with a restricted range in Illinois.

southward to extreme northeastern Mexico and Florida, fig. 93. It also includes an extension into southeastern Kansas.

SIGMODON HISPIDUS Say & Ord
Hispid Cotton Rat

The cotton rat is known no closer to Illinois than Reelfoot Lake in western Tennessee. The species is included in the key because of the possibility, although remote, that it may occur in southern Illinois.

NEOTOMA FLORIDANA (Ord)
Eastern Wood Rat

Description.—The eastern wood rat, fig. 94, is about the size of a large Norway rat. The back and head are a brownish gray or buffy gray, mixed with black, the under parts are white, and the tail is blackish above and dull white below. The eyes are large and black. The sides of the face may be grayer and the legs more brownish than the back. The ears are large and thin, and the tail is nearly as long as the body.

Length measurements: head and body 8–8¼ inches (203–209 mm.); tail 7⅜–8 inches (187–203 mm.); over-all 15¼–16¼ inches (390–412 mm.); hind foot about 1½ inches (36–40 mm.); ear from notch 1⅛ inches (28 mm.).

The skull is 39–42 mm. (about 1½ inches) long, has a long rostrum, and has high cheek teeth, each of which has a smooth grinding surface consisting of a series of triangles, fig. 74b. Dental formula: I 1/1, C 0/0, Pm 0/0, M 3/3.

Life History.—The wood rat in southern Illinois is an inhabitant of the cliffs and rocky bluffs, fig. 95, overlooking the bottomlands of the Mississippi River. It lives in the crevices and fissures within the cliffs and bluffs and among the rocky litter at the base of these. It packs quantities of materials—sticks, leaves, corncobs, cans, jar lids, empty shotgun shells, dung, or almost anything else it can carry—to the nest. Buried within this mass of rubble is the nest itself, measuring about 9 inches in diameter and composed of masses of soft material.

Little is known about the life history of the wood rat in Illinois. Fifty miles south, in Tennessee, it is known to breed in March and give birth to two or three young per litter in April.

There may be litters at other times of the year, also. This rat in Tennessee feeds on acorns, honey locust beans, beechnuts, mint, and probably many other items.

Signs.—Tracks of the wood rat are similar to those of the Norway rat. They may be found in fine dust on sheltered ledges of rocky bluffs. The tracks converge near the nest and form well-beaten pathways several inches wide. The prints of the front feet are more handlike than those of the Norway rat and have longer marks of the little toe.

An odd assortment of debris on a ledge or in a narrow crevice of a limestone cliff may indicate a wood rat house.

Fig. 94.—Eastern wood rat.

Usually there is a urinating station nearby, as well as numerous droppings, each about ¾ inch long. The nest of the wood rat is described in the section on life history.

Distribution.—The wood rat is known from very few localities in Illinois but is common where present. In this state, it has been collected only on the Mississippi River bluffs in Jackson and Union counties. The subspecies occurring in Illinois is *Neotoma floridana illinoensis* Howell. The range of the species covers an area of irregular shape, principally in the south-

Fig. 95.—Habitat of eastern wood rat in Illinois: rocky bluffs along the Mississippi River near Wolf Lake, Union County.

eastern quarter of the United States, fig. 93. It is unknown on the Eastern Piedmont Plateau.

SYNAPTOMYS COOPERI Baird

Southern Bog Lemming   Lemming Mouse

Description.—The southern bog lemming is a mouse with an exceptionally short tail and long, fine fur, which is rich brown or almost chocolate brown on the back and silvery gray on the belly. The short ears are almost concealed in the long fur.

Length measurements: head and body 4–4¾ inches (100–119 mm.); tail ⅝–¾ inch (15–21 mm.); over-all 4½–5½ inches (115–140 mm.); hind foot about ¾ inch (18–20 mm.). The ratio of hind foot to tail varies from individual to individual, but in most specimens it is 1:1. The foot may vary in length from 3 mm. shorter to 3 mm. longer than the tail.

The skull is 24.5–27.5 mm. (about 1 inch) long. It has orange-colored, broad, and grooved upper incisors, fig. 74i. Dental formula: I 1/1, C 0/0, Pm 0/0, M 3/3.

The short tail, long fur, and broad, bright orange, grooved upper incisors distinguish this mouse from all others in Illinois.

Life History.—The bog lemming occurs in wet meadows and bogs, usually where there is a thick stand of bluegrass or similar ground cover. Within this cover, bog lemmings make a series of interconnecting runways on the surface of the ground but entirely hidden beneath the vegetation. This series of miniature highways provides avenues to food supplies and to the several holes leading to underground burrows. The lemmings construct these highways by clipping away the grass with their incisor teeth and keep them smooth by constant use. The runs are no wider than the body of a lemming.

This lemming is active during daylight hours as well as at night. Most of the runs are quite dark during the daytime, for each is rather effectively roofed by a mat of grass.

The number of bog lemmings fluctuates markedly from year to year. In one year these lemmings may be overrunning a bog or meadow; in the next year few, if any, may be present. Apparently they breed from March until late fall. For rearing young, they build nests in chambers that join the underground burrows not far below the surface of the ground, or sometimes in thick mats of dry bluegrass on the surface of the ground.

The bog lemming is preyed upon by foxes, as well as hawks, owls, weasels, and most other Illinois predators. When lemmings are at the peak of their population cycle, there may be 30 or 40 per acre in suitable habitat, but, at the bottom of the cycle, there may be only 2 or 3 per acre.

Signs.—Fresh piles of grass, cut in pieces about an inch long, and bright green droppings in small, grass-covered runways may indicate the presence of bog lemmings. Only when other voles feed as exclusively on bluegrass as do lemmings, and this is rarely the case, are their droppings bright green. When there are many bog lemmings present, the greenish fecal pellets form a solid mass on the floor of much of the runway network.

Distribution.—The bog lemming, sporadic in occurrence and usually not abundant in Illinois, has been taken in only the southern two-thirds of the state. The subspecies in Illinois is *Synaptomys cooperi gossii* (Coues). The range of the species extends from eastern Quebec to southwestern Ontario, and southward to northern Arkansas and the southern Appalachian Mountains, with an extension into western Kansas.

MICROTUS PENNSYLVANICUS (Ord)

Meadow Vole   Meadow Mouse

Description.—The meadow vole, fig. 96, is a stout-bodied, short-tailed, dark-colored mouse that lives in meadow-like places. Its back is dark brown or chestnut brown; the under parts are grayish, tipped with silver (never tipped or washed with fulvous or ochraceous); the under side of the tail is nearly the same color as the belly. The hind foot usually has six pads or tubercles. The tail is nearly twice as long as the hind foot. The female has four pairs of mammary glands.

Length measurements: head and body 3⅝–4¾ inches (93–120 mm.); tail 1⅜–1⅝ inches (35–42 mm.); over-all 5–6⅜ inches (128–162 mm.); hind foot about ¾ inch (18–22 mm.).

The skull is 25–29 mm. (1–1⅛ inches) long. The front surface of each upper incisor is not grooved, and each of the last upper molars has five or six enamel triangles or loops, fig. 74*d,* not four as in the prairie vole, fig. 74*e.* Dental formula: I 1/1, C 0/0, Pm 0/0, M 3/3.

Life History.—The meadow vole lives in damp, grassy places. If the dense vegetation of such places is parted, surface

Fig. 96.—Meadow vole.

runways of this mouse may be discovered. The first runway
that is found may lead to a network of other runs. Along these
runs, voles cut off the grass; they may feed on it immediately,
or store it for future use in chambers below ground. Side runs
lead to new sources of grasses, weeds, tuberous roots, or the
bark of trees.

Underground tunnels, nearly as intricate and complex as those
aboveground, serve as refuges and breeding places for the
meadow vole. The nest of this vole is a globular mass, chiefly
of dry grass, about 5 inches in diameter. It is usually in a
chamber about 4 inches underground but it may be in a mass
of herbage aboveground.

The meadow vole can breed throughout the year. Seventeen
litters were produced by one captive female in a year. Probably
in nature, a mature female rarely produces more than six or
eight litters a year, and in some years perhaps only one or two.
A litter may consist of as many as eight young. The young are

weaned at 2 or 3 weeks of age. Because of their great fecundity, meadow voles may overrun fields during favorable years with as many as 100 or 200 individuals per acre. A disease epidemic or a food shortage may reduce the population within a year to as small a number as 2 to 10 per acre.

Signs.—Tracks, burrows, runways, and nests of the meadow vole are like those of the prairie vole. When the meadow vole walks, its footprints are paired; the prints of the hind feet fall just a little short of those of the front feet. When it runs, its footprints occur in groups of four; prints of the hind feet lie opposite one another and those of the front feet lie behind them, generally one not so far behind as the other. This pattern of footprints, and the fact that a tail mark is seldom present, may help to distinguish tracks of the meadow vole from those of the deer mouse.

Occasionally, meadow vole tracks, fig. 30, may be seen on the surface of snow, but usually voles keep to their maze of runways beneath it. The abundance and complexity of these runways show best after the snow has melted.

Piles of grass cut about an inch long and small dark brown or black droppings in surface runways are signs of the meadow vole or the prairie vole. The nest of the meadow vole is described in a paragraph on the life history of the animal.

Distribution.—The meadow vole is fairly common in extreme northern Illinois and is known to occur as far south as an imaginary line drawn between Kankakee and Havana. The subspecies in Illinois is *Microtus pennsylvanicus pennsylvanicus* (Ord). The range of the species includes all of Canada except the west coast; it extends westward into Alaska and southward in the United States as far as northern New Mexico, northern Missouri, northern Illinois, eastern Kentucky, northern Georgia, and South Carolina, fig. 93.

MICROTUS OCHROGASTER (Wagner)

Prairie Vole   Prairie Meadow Mouse

Description.—The prairie vole, fig. 4, is a short-tailed, medium-sized mouse that is common in Illinois fencerows, open grasslands, and meadows. The upper parts are a brownish gray, with a grizzled appearance resulting from a mixture of yellowish brown and black hairs. The under parts, particularly the

breast and anterior part of the belly, are washed with yellow
or rust. The tail is short, and the under side of it is somewhat
lighter in color than the upper side. Each hind foot has five pads
or tubercles. The female has three pairs of mammary glands.

Length measurements: head and body 3⅞–4⅝ inches (99–117
mm.); tail 1–1⅜ inches (26–36 mm.); over-all 4⅞–6 inches
(125–153 mm.); hind foot about ¾ inch (16–20 mm.).

The skull of this vole resembles that of the meadow vole
except that each of the last upper molars has four enamel
triangles or loops, fig. 74e, instead of six. Dental formula:
I 1/1, C 0/0, Pm 0/0, M 3/3.

The prairie vole can be distinguished from the meadow vole
by the following characteristics: rusty or yellowish wash on
under parts rather than silvery, each of the last upper molars
with four rather than five or six triangles, each hind foot with
five rather than six pads, length of tail usually 35 mm. or less
rather than over 35 mm., and three pairs of mammae rather
than four.

Life History.—The prairie vole lives in a variety of grassy
places throughout most of Illinois. In the grassy shoulders of
roads, grassy pastures that are not heavily grazed, clover fields,
grassy forest edges, and lush meadows, several prairie voles will
usually be found in each acre. Their surface runways, figs. 2
and 14, about the diameter of a garden hose, can be found by
parting the grass. These lead eventually to holes which open
into complex underground burrow systems. A small plot about
16 feet square may have as many as 160 feet of surface runs
and 10 to 40 feet of underground burrows. The prairie vole
obtains grasses, seeds, and various herbs for food from along
surface runways. It finds refuge, feeds on roots, and nests in
the underground burrows.

Storage chambers for food, fig. 2, and usually the nest of the
prairie vole are located in an enlarged portion of an underground burrow. The nest consists of grass—in some cases the
equivalent of several pints of it. If the grass in which this vole
lives is very dense and tall, the nest may be placed aboveground
among the roots of the grass. Young are produced throughout
most of the year except in the coldest winter months. There
are three to six young in each litter. The young develop as rapidly as do those of the meadow vole. A female of the prairie
vole is sexually mature when only 3 or 4 weeks old.

The prairie vole may be active day and night, winter and summer, but is probably most active at night. In winter it makes burrows through the snow at the surface of the ground. Under the protective mantle of snow, it may make runs to the trunks of trees, where it feeds on the bark.

When conditions are ideal, prairie voles may increase rapidly, and a population peak of 100 or more per acre may result. Usually within a year or two after the peak has been reached, there will be a rather rapid and marked decrease in the number of voles. Later, the population may build up to another peak. Increases and decreases in population follow a rather definite time pattern or cycle. Biologists believe that the cycle is completed about every 4 years.

Signs.—The runways and nests described above are signs of the presence of the prairie vole. Other signs are tooth marks, beginning immediately above ground, on the trunks of trees, fig. 37. Tracks of the prairie vole are like those of the meadow vole.

Distribution.—The prairie vole is most abundant in central and southern Illinois but is known to occur in suitable habitats throughout the state. The Illinois subspecies is *Microtus ochrogaster ochrogaster* (Wagner). The range of the species covers an area from western West Virginia to southeastern Alberta and southward to central Oklahoma and western Tennessee.

PITYMYS PINETORUM (Le Conte)

Pine Vole   Pine Mouse

Description.—The pine vole, fig. 97, is a short-tailed mouse with small eyes. The fur on the back and sides is chestnut or bright brown, with a gloss or sheen, and is dense, almost like that of a mole. The under parts are gray and may be washed with buff. The tail is brown, lighter on the under than on the upper side. The hind foot is usually 1 to 3 mm. shorter than the tail. There are four (two pairs of) mammary glands.

Length measurements: head and body 3¾–4¼ inches (95–108 mm.); tail about ¾ inch (15–22 mm.); over-all 4¼–5⅛ inches (110–130 mm.); hind foot about ⅝ inch (16–18 mm.).

The skull is much like that of other voles but is broader through the interorbital region (more than 4 mm., or more than ⅛ inch) and the posterior border of the palate has a

Fig. 97.—Pine vole.

median projection. Dental formula: I 1/1, C 0/0, Pm 0/0, M 3/3.

The pine vole, fig. 4, differs from other voles in having a shorter tail, glossy, velvety fur, two pairs of mammary glands, and a broad interorbital region of the skull. It differs from the bog lemming in possessing ungrooved and narrower upper incisors and glossy brown hair that is velvety like that of a mole.

Life History.—The pine vole occurs in woods, orchards, and even in grassy fields some distance from woods. It lives in leaf litter or grassy mats. It makes an extensive network of underground burrows, which it uses in searching for food, in resting or nesting, and in rearing its young. Along these underground burrows, fig. 1, it feeds on tubers and succulent roots, grasses, seeds, and roots of trees. The pine vole spends much less time in surface runways than do other Illinois voles. It makes a globular nest of dead leaves and grasses in a burrow a few inches below the surface of the ground, under the roots or the stump of a tree, or in a log.

Breeding of the pine vole extends from March to November. Usually there are three or four young in a litter. Although helpless at birth, the young can gather food for themselves at about 2 weeks of age.

The pine vole is preyed upon by foxes, minks, skunks, raccoons, and owls.

Signs.—Tracks of the pine vole are similar to those of other voles but are rarely seen because this mouse prefers to remain underground. Its burrows, however, are common in some woodlands and orchards. These burrows may be just beneath a thick carpet of leaves or deeper in the soil. They vary in diameter from somewhat less than 1½ inches to as much as 2 inches. Some of them may surface beneath well-rotted logs. In orchards they may lead to fallen apples, on which the vole may feed from beneath. Pine vole burrows are often appropriated by shrews.

Distribution.—The pine vole is state-wide but sporadic in occurrence and usually uncommon. There are two subspecies in Illinois, *Pitymys pinetorum auricularis* (Bailey) in the southern third of the state and *P. p. scalopsoides* (Audubon & Bachman) elsewhere. The species occurs from Massachusetts and Vermont to central Wisconsin and south to central Texas and northern Florida.

ONDATRA ZIBETHICUS (Linnaeus)

Muskrat

Description.—The muskrat, fig. 98, is a large vole, nearly the size of a cottontail rabbit, that is adapted to a life in and near water. Its body is a dark or chocolate brown, darkest on the back where long glistening guard hairs are thickest. The eyes and ears are small. The hind feet are large and webbed. The blackish, nearly naked tail is laterally flattened, long, and scaly.

Length measurements: head and body 11–12¾ inches (280–325 mm.); tail 8¼–10¾ inches (210–275 mm.); over-all 19¼–23½ inches (490–600 mm.); hind foot 2⅞–3⅜ inches (73–85 mm.). Weight: about 2½ pounds.

The skull is more than 60 mm. (2⅜ inches) long; the grinding or cheek teeth, fig. 40c, are long and flat surfaced, each consisting of a series of upright prisms or columns; the incisors are not grooved. Dental formula: I 1/1, C 0/0, Pm 0/0, M 3/3.

Life History.—The muskrat lives along or in the many rivers, streams, drainage ditches, marshes, lakes, ponds, and water-filled strip mine areas of Illinois. Usable habitat for the muskrat has been increased by construction of drainage ditches and strip mine areas and by protection of stream and ditchbanks from overgrazing and erosion.

The muskrat is at home in water, on land, and below the surface of the ground. Muskrats that live in marshes, ponds, and strip mine areas usually build houses, figs. 3 and 11, some for

Fig. 98.—Muskrat.

dwelling and some for feeding, from materials that are readily available, such as bulrushes, smartweed, and cattails. A dwelling house usually is dome shaped, several feet in diameter at the base, and the walls are 1 to 2 feet thick. The entrances (usually two) are below the surface of the water. Within the house the muskrat makes its nest, which is above the normal high-water level and usually remains dry. Perhaps half or more of the houses are feeding houses. These are without nests, are smaller than the dwelling houses, and have thinner walls.

Muskrats that live along rivers and ditches build no houses like those described above but extend burrows back into the banks. The nest chambers are above the surface of the water at its normally highest level, and thus the nests usually remain dry. Entrances are normally below the surface of the water, but at times of low water some of them may be exposed. Runways or paths lead from the exposed burrow entrances to the water. Trails visible in shallow parts of a stream may indicate usual routes of travel.

The muskrat may breed from April to September. A female usually has two litters, occasionally three, per year, and litters average about four young each. At about a month of age, the young are sufficiently grown to shift for themselves. Sometimes there may be room within the parental marsh or along the home bank for these new muskrats to settle down, for an acre of suitable marsh may accommodate 20 to 40 rats, or a mile of ungrazed ditchbank more than a hundred. At other times, some of the young may have to move long distances to establish homes, and it is during forced migrations overland that many are killed by motor vehicles on highways or by dogs or other predators. Once established in their new homes, the young muskrats may be preyed upon by minks. Extreme fluctuations in water level can be destructive to an entire population of muskrats.

The muskrat is active during all months of the year. It feeds on roots, tubers, and green material, including cattails, bulrushes, sedges, pickerelweeds, corn, alfalfa, wild parsnips, willows, and clovers.

Signs.—Houses in ponds or marshes, fig. 11, runways from water to burrows in the banks of streams or ditches, fig. 12, and feeding platforms among the vegetation in shallow water, fig. 38, are telltale signs of muskrats. Footprints and tail marks, fig. 21, may be evident at the water's edge. The print of a hind foot is about 3¼ inches long; the print of a front foot shows only four toes and is about 1¼ inches long. Droppings, fig. 21, are dark brown, oval-shaped, and each a little more than ½ inch long; they are frequently deposited on stones and logs projecting above water.

Distribution.—The muskrat, found in suitable habitats throughout Illinois, is most common in the northern portion of the state. The Illinois subspecies is *Ondatra zibethicus zibethicus* (Linnaeus). The range of the species includes most of Canada, Alaska, and the United States. It does not include south-central Oregon, most of California, south-central Nevada, southern Arizona, south-central Texas, or the extreme southeastern United States.

RATTUS RATTUS (Linnaeus)

Roof Rat   Black Rat

Description.—The roof rat, fig. 99, generally is a grayish brown on the upper parts and a creamy color on the under parts. The tail is brown all around and longer than the body. The upper sides of the feet are light brown. Occasionally black individuals or families occur. This rat differs from the Norway rat in that it has a slimmer body and a more uniformly colored tail that is longer, rather than shorter, than its head and body. Its ears are larger, more delicate, and less hairy.

Length measurements of one specimen from Illinois: head and body 6¾ inches (171 mm.); tail 10½ inches (266 mm.); over-all 17¼ inches (437 mm.); hind foot 1½ inches (37 mm.).

The skull has the temporal ridges on the lateral margins of the braincase bowed slightly outward, and the length of a parietal measured along a temporal ridge is noticeably less than the distance between these ridges; otherwise the skull is like that of the Norway rat. Dental formula: I 1/1, C 0/0, Pm 0/0, M 3/3.

Life History.—In many places in the United States the roof rat has been driven out of favored habitats by its larger cousin,

Fig. 99.—Roof rat.

the Norway rat. As a result it often makes its home in the upper parts of buildings and in trees, for it is a better climber than the Norway rat. It is slightly less prolific.

Distribution.—In Illinois, the roof rat is known from one specimen taken in Urbana, but it may occur from time to time in other cities over the state through accidental introductions from the South. The one known Illinois specimen, a black-haired individual, belongs to the subspecies or variety *Rattus rattus rattus* (Linnaeus). The roof rat, introduced from the Old World, commonly occurs along the Atlantic and Gulf Coast states from New York to Texas and southwest throughout central Mexico. It occurs also along the Pacific Coast from southern British Columbia southward into Mexico and occasionally in interior parts of the United States. A colony

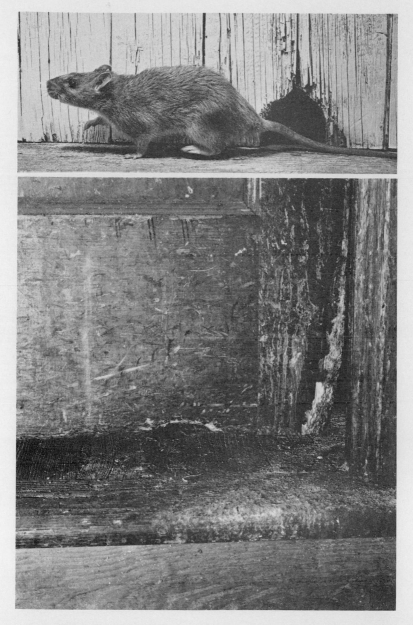

Fig. 100.—Top, Norway rat; bottom, door that has been gnawed by a Norway rat.

of the brown variety, *R. r. alexandrinus* (E. Geoffroy-Saint-Hilaire), inhabited some warehouses at St. Louis, Missouri, in 1945 but it was soon exterminated.

RATTUS NORVEGICUS (Berkenhout)

Norway Rat   Barn Rat   House Rat   Sewer Rat

Description.—The Norway rat, fig. 100, generally is grayish brown on the upper parts and gray on the under parts. The fur is coarse. The gray or brown tail is slightly but noticeably lighter below than above, is scaly, nearly devoid of hair, and slightly shorter than the head and body.

Length measurements: head and body 7¼–9¾ inches (187–250 mm.) ; tail 5½–7¼ inches (138–185 mm.) ; over-all 13–17 inches (325–435 mm.) ; hind foot 1½–1¾ inches (38–44 mm.). Weight: average about ⅔ pound, but there is much individual variation.

The skull is 41.5–45.0 mm. (nearly 1¾ inches) long; the rostrum is long; ridges may be present on the lateral margins of the braincase; the cusps on the cheek teeth of the upper jaw are in three longitudinal rows, fig. 74c. Dental formula: I 1/1, C 0/0, Pm 0/0, M 3/3.

Life History.—The Norway rat, an Old World species accidentally introduced into the New World, is at home not only in dwellings, warehouses, stores, and sewers of American cities but around garbage dumps, along ditchbanks, under straw piles, and in cornfields, feed lots, and barns. Usually it chooses a protective shelter from which it can burrow into the ground, fig. 8. It is a sociable animal, and usually several of its kind live together.

This rat reproduces at a rapid rate. Litters are made up of 2 to 15 young and average about 7. The young are weaned at 3 weeks and are capable of breeding when about 3 months old. The gestation period is 21 to 23 days. Litters may be produced almost continuously throughout the year if adequate food is available. Theoretically, one pair of Norway rats and their progeny can produce more than 3,000 rats in a single year.

In some years, Norway rats become excessively abundant. In 1903, hordes of rats were found migrating over several counties in western Illinois. In 1939, rats were again very abundant in this state.

The Norway rat is truly omnivorous. It eats the kinds of foods consumed by man and many other things besides. An adult rat usually requires a minimum of ¾ ounce of dry food and ½ to 1 ounce of water daily.

This rat is able to enter buildings by climbing wires or faces of rough walls, by gnawing through wood, or by burrowing under masonry foundations.

In most years the citizens of Illinois are the custodians of about one Norway rat per person. A rat that lives near supplies of food for man or livestock eats, fouls, or otherwise wastes more than 20 pounds of food or feed each year. Also it undermines foundations and damages wooden structures by its burrowing and gnawing.

In the wild, the Norway rat is preyed upon by foxes, weasels, minks, owls, hawks, snakes, and other animals. In urban districts and around farm buildings, man can be its worst enemy.

Signs.—The Norway rat makes tracks, fig. 22, that are like those of a tree squirrel, but the prints of the front feet are smaller and less often paired; also, a tail mark often shows. Droppings are oblong and usually up to ¾ inch long. A dark smudge along a baseboard molding or around a hole in a wall or floor is indicative of a frequently used path of a rat. A burrow of this rat has piles of dirt and often corncobs or pieces of junk at the entrance. Gnawed doorways, fig. 100, are common in poorly sanitated sections of cities.

Distribution.—The Norway rat is abundant throughout Illinois. Only the subspecies *Rattus norvegicus norvegicus* (Berkenhout) is known in the United States. The Norway rat occurs in southern Canada, along the Pacific Coast to northern Alaska, and throughout the United States and Mexico.

MUS MUSCULUS Linnaeus

House Mouse

Description.—The house mouse, fig. 101, is grayish brown to dark gray on the upper parts and gray or buff on the under parts. Its practically hairless tail is uniformly brownish gray.

Albinos of this species are the "white mice" of laboratories.

Length measurements: head and body 2¾–3⅜ inches (70–85 mm.); tail 2⅜–3¾ inches (60–95 mm.); over-all 5–7 inches (130–180 mm.); hind foot ⅝–¾ inch (16–19 mm.).

The skull is readily recognizable by a distinct notch in the tip of each upper incisor as viewed from the side, fig. 74*f*, and by three longitudinal rows of cusps in each of the upper cheek teeth. Dental formula: I 1/1, C 0/0, Pm 0/0, M 3/3.

This is the only mouse in Illinois that has gray or buffy under parts, a long, naked tail that is almost the same color all around, and a peculiar notch at the tip of each upper incisor.

Life History.—The house mouse, like the Norway rat, is an accidentally introduced pest. It lives in dwellings, office buildings, stores, factories, barns, poultry houses, and sheds, and in the runways of native mice and voles. Frequently, mice of this species move out of houses into yards, gardens, and fields in the spring and back into houses in the fall. They usually dominate native mice, and may forcibly drive them out of an area.

The house mouse produces litters throughout the year, except for the cold winter months. The female has five or six young per litter. The young can run about at 21 days of age and breed at 42 days.

In eating habits, this mouse is truly omnivorous.

Signs.—Gnawed paper, black droppings about ¼ inch long, or small tracks in the dust beside a building may be evidence of the presence of this mouse. In fields, it is difficult to distinguish between signs of the house mouse, fig. 32, and those of some of the native mice.

Distribution.—The house mouse, like the Norway rat, is abundant throughout southern Canada and along the Pacific Coast to northern Alaska, the United States, and Mexico. The

Fig. 101.—House mouse.

subspecies in Illinois is probably *Mus musculus domesticus* Rutty.

ZAPUS HUDSONIUS (Zimmermann)
Meadow Jumping Mouse

Description.—The meadow jumping mouse, fig. 102, has large hind feet and a tail that is considerably longer than the body. It has no cheek pouches. The fur of this animal is dark olive-brown on top of the back, finely streaked with dark brown hairs on the sides, and white on the under parts.

Length measurements: head and body about 3 inches (about 77 mm.); tail 4¼–5¼ inches (108–135 mm.); over-all 7¼–8¼ inches (185–212 mm.); hind foot 1–1¼ inches (25–31 mm.); ear from notch about ½ inch (11.0–15.5 mm.).

Fig. 102.—Meadow jumping mouse.

The skull is about 22 mm. (⅞ inch) long and has a large infraorbital foramen. There are four cheek teeth on each side of the upper jaw. The upper incisors are grooved. Dental formula: I 1/1, C 0/0, Pm 1/0, M 3/3.

Life History.—The meadow jumping mouse may be found at night on a grassy or vine-covered bank of a stream or pond, usually where there are a few trees. It seldom runs but travels rapidly by making a series of jumps, each nearly a yard in length. If pursued, it may take refuge among roots and debris or it may dive into water without hesitation. It swims well and dives frequently to avoid capture, but in water it tires rather quickly.

This mouse makes its home in a globular nest of grass, other herbs, or moss. The female gives birth to young between May and August. This mouse feeds most extensively on grass seeds, berries, nuts, and insects. The stored fat in the body prepares

it for the winter sleep. It hibernates, probably from November
until early April, far enough underground to be well below
the frost line.

Signs.—Because it is in hibernation when snow is on the
ground, and because it is too light to make visible tracks in mud,
the meadow jumping mouse seldom leaves tracks that can be
detected.

In areas where this mouse is common, little crisscross piles of
slender grass and weed stems may be found; the stems have been
cut into 2- to 3-inch sections in attempts of the animal to reach
the seed heads. The piles are as much as 6 or 7 inches across.
The sections of stems may be in little bundles and are not to be
mistaken for the shorter sections, averaging about 1½ inches
long, left by the meadow vole.

Summer nests of this mouse are in open meadows: on the
surface, in a tussock of grass, or a few inches below the surface
of the ground, covered by a protecting log. They are made of
material found nearby, often entirely of the dead leaves of
herbs, bits of grass, or dry moss.

Distribution.—The meadow jumping mouse, although not
abundant, is of state-wide occurrence in Illinois. The subspecies in this state is *Zapus hudsonius intermedius* Krutzsch.
The range of the species extends westward from Labrador
across most of Canada and Alaska, and southward to northeastern Colorado, northern Oklahoma, northeastern Georgia, and
northern South Carolina.

ORDER LAGOMORPHA

Hares and Rabbits

Members of the order Lagomorpha—hares, rabbits, and
pikas—resemble the rodents in having a pair of large, chisel-
shaped upper incisor teeth, but differ from them in having an
additional small, inconspicuous incisor directly behind each of
these teeth, fig. 40*d*. They also differ from the rodents in having
short tails that, in most species, resemble small tufts of cotton.
All lagomorphs are herbivorous.

Only three species of lagomorphs, all of which belong in the
family Leporidae, occur in Illinois. Of these, the eastern cottontail is the most abundant and widely distributed. It is probably the most important game animal in the state, being taken

by more than three-fourths of the hunters and making up nearly
half of the total bag of game. The other two species are restricted in range and are of little importance from a hunter's
standpoint.

Economic Status.—In some years cottontail rabbits do considerable damage in orchards and nurseries in Illinois by eating
the bark of young trees, fig. 37. Also, at times they are trouble-
some in truck and home gardens.

Rabbits are the source of most cases of human tularemia in
the United States. In 1939, in a period of particularly heavy
outbreaks among rabbits, 485 human cases of this serious disease were reported in Illinois. In most years since then, the
number of human cases in this state has been less than 100 annually. More than half of the cases in Illinois in the 1936–1949
period were from the southern third of the state (Yeatter &
Thompson (1952:357). The tick that is the chief vector of this
disease feeds on its rabbit host from about the end of winter
until after the return of freezing weather in autumn, when the
tick becomes immobile and drops off its host. In about a week
after the onset of freezing weather, most of the infected rabbits
will have died and any rabbits not then infected will be free of
the disease for about 2 months. Sickly or sluggish rabbits and
rabbits found dead should be avoided as possible sources of the
disease.

The domestic rabbit, *Oryctolagus cuniculus,* has been included
in the key for the identification of skulls because occasionally a
skull of this animal may be found. Not only have individuals
of this species been known to escape from captivity, but efforts
have been made to establish the species as a game animal in
Illinois. The domestic species and native species do not cross.

KEY TO SPECIES

Whole Animals

1. Ear from notch more than 75 mm. (3 in.) long; hind foot
     more than 120 mm. (4¾ in.) long . . . .
     . . . . .white-tailed jackrabbit, *Lepus townsendii*
     Ear from notch less than 75 mm. long; hind foot less than
     120 mm. long 2
2. Hind foot less than 90 mm. (3½ in.) long; back of animal
     grayish brown . . . . .eastern cottontail, *Sylvilagus floridanus*
     Hind foot more than 90 mm. long; back of animal reddish,
     with much black . . . . .swamp rabbit, *Sylvilagus aquaticus*

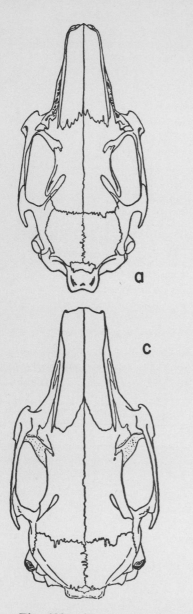
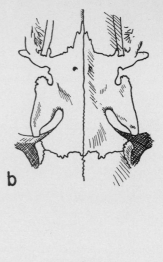
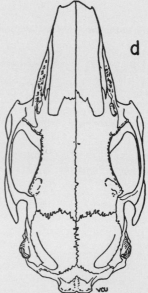

Fig. 103.—Characters used in the identification of lagomorphs:
a, skull of domestic rabbit, top view; *b,* portion of skull of jackrabbit, top view; *c,* skull of cottontail, top view; *d,* skull of swamp
rabbit, top view.

Skulls

1. Supraorbital processes each with the posterior projection
well separated from the braincase, fig. 103*a, b*; base of
skull fenestrated 2
Supraorbital processes each with the posterior projection or
extension partly or completely fused to braincase, fig. 103*c,
d*; base of skull not fenestrated....................... 3
2. Nasals gradually narrowed from base to near apex; posterior projection of each supraorbital process narrow, fig.
103*a*..................domestic rabbit, *Oryctolagus cuniculus*
Nasals almost parallel from base to near apex; posterior projection of each supraorbital process broad, fig. 103*b*......
.... white-tailed jackrabbit, *Lepus townsendii*
3. Each supraorbital process with part of posterior projection
fused to braincase, leaving a distinct slitlike opening,
fig. 103*c*...........eastern cottontail, *Sylvilagus floridanus*
Each supraorbital process with posterior extension almost
entirely fused to braincase, leaving only a small pore, fig.
103*d*..................swamp rabbit, *Sylvilagus aquaticus*

LEPUS TOWNSENDII Bachman
White-Tailed Jackrabbit

Description.—The white-tailed jackrabbit, fig. 104, is much
larger than either the eastern cottontail or the swamp rabbit.
The upper parts in summer are buffy or brownish gray, in
winter white; the under parts are white, except for the throat,
which is buff. The ears are long and tipped with black. The
tail is white. The hind legs are long, and the feet are well
furred.

Length measurements: head and body 19–20 inches (475–505
mm.); tail 4–4½ inches (100–115 mm.); over-all about 24
inches (575–620 mm.); hind foot 5⅜–6 inches (138–153 mm.);
ear from notch 4–4½ inches or about 3¾ inches dry (96–113
mm.). Average weight: about 7 pounds.

The skull is about 95 mm. (3¾ inches) long; the rostrum is
greatly fenestrated. The bullae are comparatively large; the
interparietal is fused with the parietals (unfused in *Sylvilagus*);
and the posterior projection of each supraorbital process is free
(not fused to the braincase), fig. 103*b*. Dental formula: I 2/1,
C 0/0, Pm 3/2, M 3/3.

Life History.—The white-tailed jackrabbit makes its home
in open country, where with its large ears it can hear approach-
ing trouble from a considerable distance and with its long legs

Fig. 104.—White-tailed jackrabbit in winter pelage.

it can usually outrun its enemies. Many a dog has pursued a white-tailed jackrabbit only to become exhausted while the hare was still fresh and several hundred feet in the lead.

This animal rests in forms, which are body-sized depressions either in thick vegetation or in soft soil. Usually it feeds on grasses, clovers, and other herbs, and on grains. In winter, it may be forced to feed on buds, bark, and twigs of woody plants.

The white-tailed jackrabbit is really a hare and not a true rabbit. Its precocious young are fully furred and have their eyes wide open at the time of birth. Probably it breeds in April, and young, three to six in a litter, are born in June. The young may nibble on grasses when only a few days old and in less than 5 or 6 weeks are ready to shift for themselves.

Signs.—Prints left by the hind feet of a jackrabbit, fig. 16c, are much larger than those of the cottontail, the length of one print varying from 4 to 6¼ inches as compared with 4 inches or less for that of the cottontail. The tracks of the front feet are also correspondingly larger than those of the cottontail. The patterns of the tracks of the two animals are much alike. The long prints of the hind feet lie approximately side by side and

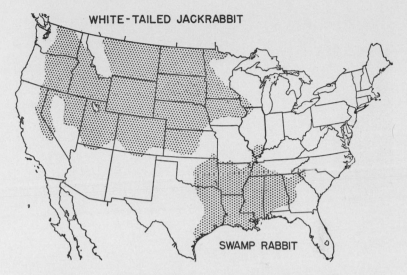

Fig. 105.—Known distribution, in the United States, of two lago-morphs with a restricted range in Illinois.

a little ahead of the small prints of the front feet, one of which trails the other when the animal hops, fig. 16c.

Droppings of the jackrabbit average larger than those of the cottontail but are otherwise like them.

Jackrabbit forms, or resting places, are about 15 inches long and half as wide; cottontail forms are about 10 inches long. A form found in a plowed field, fig. 10, almost certainly belongs to a jack rather than a cottontail or a swamp rabbit.

Distribution.—The white-tailed jackrabbit occurs in Illinois only in the extreme northwestern counties. It is common in the sand prairie at the Savanna Ordnance Depot in Jo Daviess County. The subspecies in Illinois is *Lepus townsendii cam-panius* Hollister. The range of the species extends from north-western Illinois north and west to central Saskatchewan and central Oregon and south to east-central California and north-ern New Mexico, fig. 105.

SYLVILAGUS FLORIDANUS (Allen)
Eastern Cottontail

Description.—The eastern cottontail, frontispiece and fig. 106, is the common rabbit of Illinois. The upper parts of the

body are buff or rusty brown, and the under parts, except for
the buff throat, are white. The ears are darker than the back,
the nape of the neck is reddish, and the under side of the tail
is white.

Length measurements: head and body 13¾–15½ inches (350–
395 mm.); tail 1¾–2½ inches (45–65 mm.); over-all 15½–18
inches (395–460 mm.); hind foot 3⅜–4¼ inches (85–110 mm.);
ear from notch, about 2⅛ inches (50–60 mm.). Weight: usually 2–3 pounds.

The skull, figs. 40*d*, 103*c*, is about 70 mm. (2¾ inches) long
and, except for the distinct interparietal bone, is not unlike that
of the jackrabbit. Other characters are noted in the key to
skulls. Dental formula: I 2/1, C 0/0, Pm 3/2, M 3/3.

Fig. 106.—Young of eastern cottontail.

Life History.—The eastern cottontail makes its home in
brushy or weedy fields, fig. 2, thickets along fencerows and margins of wood lots, forest edges, and dry bottomlands. It is uncommon in pastured woodland, fig. 116. It frequently forages
in intensively cultivated fields that are situated near permanent
cover. At dawn and at dusk it is often seen feeding in such
fields and in open grassy areas.

In central Illinois, the male reaches breeding condition in
mid-February, the female in late February. The breeding season attains its peak during the first 2 weeks of March and continues into September. Usually the female produces annually
three litters of five or six young each. The nest is placed in a
depression in the soil. It has an outer lining of grasses and
leaves and an inner lining of hair, which the female has pulled
from her breast and abdomen. The nest is completely covered
while the young are alone in it. At birth the young have short
fur and are blind but capable of creeping about. In about a

week, their fur is considerably longer and their eyes are open. At about 2 weeks of age the young nibble on grasses.

The cottontail feeds preferably on grasses and broad-leaved weeds, but eats other vegetation also. It commonly feeds on bluegrass, clover, dandelion, plantain, alfalfa, and soybean plants. In winter, it may be forced to turn to buds, bark, and twigs, particularly of dogwood, elm, rose, and apple.

A multitude of enemies harass the cottontail. Among these are dogs, foxes, owls, weasels, and minks, as well as men and their guns. The cottontail attempts to avoid predators by running away or by freezing. It will remain quiet and still as if frozen to a spot until danger approaches too close; then, with an explosive burst of speed, it will bounce to another spot sufficiently far away, and where cover is available, and again freeze.

Signs.—Footprints of the cottontail in snow, dusty paths, or soft damp ground are easily recognized. When this animal travels at its usual gait, which is a hop, prints of the hind feet lie side by side and in front of those of the front feet, one of which trails the other, fig. 16. The print of each hind foot is 2 to 4 inches long. When the rabbit is hopping slowly, the prints of the front and hind feet may lie side by side but, when it is feeding, the front feet are placed in front of the hind feet. Each print is pointed like the tip of an arrowhead. The spherical droppings are each about 10 mm., or somewhat less than ½ inch, in diameter.

Distribution.—The eastern cottontail, common throughout Illinois, fluctuates in abundance from year to year. Two subspecies occur in this state, *Sylvilagus floridanus alacer* (Bangs) south of the Shawnee Hills and *S. f. mearnsii* (Allen) in the rest of Illinois. The range of the species includes the eastern United States as far north as northeastern Connecticut and central New York, westward through southern Ontario to southern Manitoba, and southward through parts of Colorado, Texas, New Mexico, and Arizona into Mexico.

SYLVILAGUS AQUATICUS (Bachman)
Swamp Rabbit

Description.—The swamp rabbit is similar to the eastern cottontail, but it is larger, is darker on the upper parts, and has more reddish brown on the body.

Length measurements: head and body 18–18½ inches (455–468 mm.); tail 2½–2¾ inches (65–72 mm.); over-all 20½–21¼ inches (520–540 mm.); hind foot about 4¼ inches (105–112 mm.); ear from notch about 2½ inches dry (63–68 mm.). Weight: 3–5½ pounds.

The skull, about 85 mm. (3⅜ inches) long, has a minute hole between the postorbital process and frontal bone on each side, fig. 103d. Dental formula: I 2/1, C 0/0, Pm 3/2, M 3/3.

Life History.—In southern Illinois the swamp rabbit lives in cane thickets or dense woods and brush bordering swamps. It is never far from water, and, being a good swimmer, does not hesitate to take to water. It swims with little more than the top of its head exposed and with its ears straight back, fig. 3. The denseness of the thickets in which it lives and its ability to swim are valuable protective features.

This rabbit feeds on a variety of green herbs, bark, and leaves; its fondness for giant cane is indicated by one of its popular names, "cane-cutter." Its nest resembles that of the cottontail, but it is larger and frequently is in thickets with stalks pulled down to form a protective covering. The young are produced between March and November, usually four at a time; a female may have more than one litter in a year. At birth, the young have very short fur and are blind.

Signs.—Footprints of the swamp rabbit are similar to but usually larger than those of the cottontail (print of hind foot 3–4½ inches). Droppings are like those of the cottontail but are deposited on logs or mounds near water, fig. 3.

Distribution.—The swamp rabbit occurs in the southern third of Illinois. The subspecies in this state is *Sylvilagus aquaticus aquaticus* (Bachman). The range of the species extends from northwestern South Carolina to eastern Texas and includes two narrow northward projections, one into southern Illinois and southwestern Indiana and the other into southeastern Kansas, fig. 105.

ORDER PERISSODACTYLA
Odd-Toed Hoofed Mammals

Members of the order Perissodactyla have on each foot an odd number of hoofed toes; the middle (third) toe is distinctly larger than the others and forms the principal support of the foot. The order includes the horse, tapir, and the rhinoceros, each known from only a few species. There are no living native North American representatives of the order in Illinois, but the domestic horse, of Old World origin, is familiar to all.

The horse is mentioned in this Fieldbook not because anyone would have any difficulty in identifying the whole animal, but because a collector might find a horse skull or tooth, which is not readily identified. The skull and molar tooth of a horse are illustrated in figs. 40f and 109c, and identifying features are given in the key to orders, beginning on page 41.

ORDER ARTIODACTYLA
Even-Toed Hoofed Mammals

Mammals of the order Artiodactyla have a large, hard hoof on the third and another on the fourth toe of each foot, and the principal support for the foot is shared by these two toes. The result is the forming of the familiar so-called "cloven hoof." The other toes may be represented by smaller hoofs or by vestigial structures. The order is a large one, representatives of which are distributed all over the world. It contains many strange animals, such as the hippopotamus, camel, and giraffe. North American representatives include members of the deer (Cervidae), sheep and bovine (Bovidae), pronghorn (Antilocapridae), and pig (Suidae) familes.

When white men first arrived in the Illinois country, three species of these artiodactyls occurred wild in the area—the bison, the elk or wapiti, and the white-tailed deer. Land settlement and intensive farming caused all three to disappear from Illinois in the previous century, but in recent years the white-tailed deer has been successfully reintroduced. This species is the only wild representative of the order Artiodactyla in the state.

Several domestic animals belong to this order—the cow, sheep, goat, and pig. These are well known, and no key is needed for identification of the whole animal. There is a good possibility, however, that if skulls of these animals, figs. 107a, 108a, b, c, are found they will be confused with deer, elk, or bison skulls that may be unearthed. A key has been added for the identification of these skulls. Some of the molar teeth of the cow are illustrated in fig. 109a, b.

KEY FOR SKULLS OF THE ARTIODACTYLA

(Domestic species included)

1. Each eye socket incompletely encircled by bone, fig. 107*a*;
    upper jaw with incisors...................pig, *Sus scrofa*
    Each eye socket completely encircled by bone, fig. 107*b*; up-
    per jaw without incisors............................... 2
2. Upper jaw with canine teeth, fig. 107*b*................... 9
    Upper jaw without canine teeth........................ 3
3. Skull more than 12¼ inches (313 mm.) long........... 4
    Skull less than 12¼ inches long....................... 6
4. Open space present between lacrymal and nasal bones, fig.
    107*b*...............................elk, *Cervus canadensis*

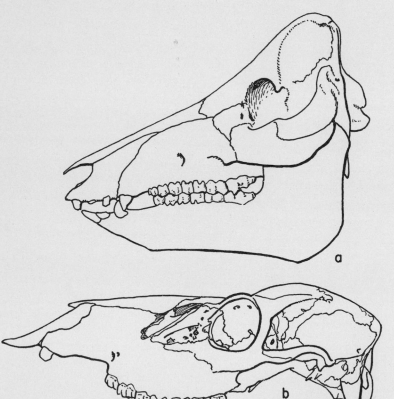

Fig. 107.—Characters used in the identification of ungulates: *a*, skull of pig, side view; *b*, skull of American elk, side view, but without lower jaw.

No open space between lacrymal and nasal bones, fig.
108*b, d* 5

5. Skull elongate, fig. 108*b*; prominent ridge at top of skull between horns or horn bases..............................
.......................domestic cow, bull, or ox, *Bos taurus*

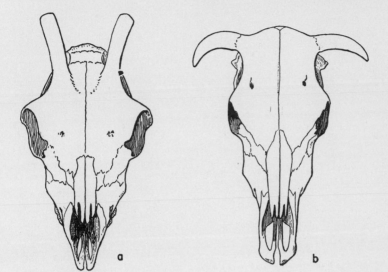

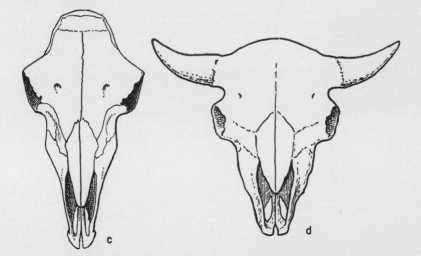

Fig. 108.—Dorsal views of skulls of ungulates: *a,* goat; *b,* cow;
c, sheep; *d,* bison.

Skull subtriangular in frontal view, fig. 108*d*; no prominent
ridge between horns or horn bases....... bison, *Bison bison*
6. Top of skull with a groove extending from the inner side of
the base of each horn (or knob) and meeting on the forehead to form a V or U, fig. 108*a*; horns usually present,
subparallel, and directed posteriorly........goat, *Capra* sp.
Top of skull without such grooves, fig. 108*c*; horns absent,
or diverging and downwardly curved................... 7
7. Skull relatively long and narrow; posterior narial aperture
completely divided by a vertical vomerine partition; antlers present or absent ....
..................white-tailed deer, *Odocoileus virginianus*
Skull relatively short and wide; posterior narial aperture
without complete vertical septum; antlers never present.. 8
8. Skull with sides behind eyes converging sharply, fig. 108*c*;
bony eye sockets strongly protuberant.....................
....................................domestic sheep, *Ovis* sp.
Skull with sides behind eyes rounded; bony eye sockets not
strongly protuberant..............domestic calf, *Bos taurus*
9. Skull more than 12¼ inches long (313 mm.)............
.... elk, *Cervus canadensis*
Skull less than 12¼ inches long........................
..................white-tailed deer, *Odocoileus virginianus*

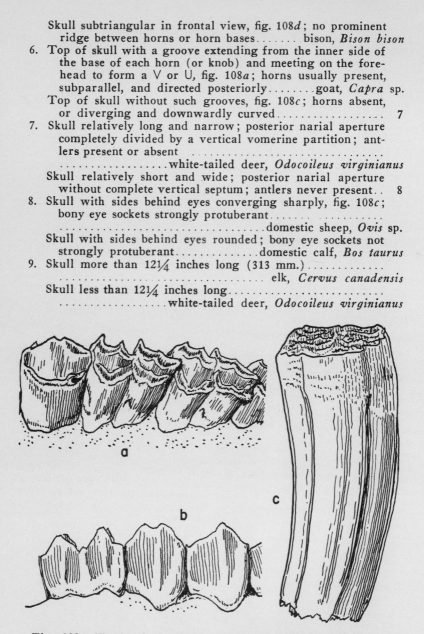

Fig. 109.—Teeth of ungulates: *a,* grinding surface of teeth of
cow; *b,* outer side of teeth of cow; *c,* tooth of horse.

CERVUS CANADENSIS Erxleben

Elk   Wapiti

Description.—The elk is the largest member of the deer family ever known to occur in Illinois. It weighs about 650 pounds (females smaller than males) and stands nearly 5 feet high at the withers (shoulders). The back and sides of the body are grayish brown, and a large patch on the rump is buff or yellow. An old male may have antlers several feet long and nearly 2 inches in diameter at the base. The branches of antlers are directed forward. The two canines are knoblike, fig. 107b.

Former Distribution.—The elk was once found along the wooded streams and in the open woods of Illinois. Some authorities believe that it was at home on the prairies also. Animals of this species are gregarious; they gather in large herds in winter, small herds in summer. They browse on twigs and leaves of trees and graze on many different kinds of green plants.

The elk or wapiti once occurred over most of Illinois. By the early 1800's it was becoming uncommon, and by the mid-1800's it had disappeared from the state. Available Illinois records for the elk are for Cook County in the northeastern, Kaskaskia in the southwestern, near Peoria in the central, and near Mount Carmel in the southeastern parts of the state.

ODOCOILEUS VIRGINIANUS (Zimmermann)

White-Tailed Deer   Virginia Deer

Description.—The white-tailed deer, figs. 110 and 111, in Illinois can be confused with no other animal. Its large size, long legs, thick, stubby but conspicuous white tail, and ornate antlers in males distinguish it from other mammals. In winter, the coat is gray or grayish brown, in summer reddish brown. The under parts of the body and the under side of the tail are white. The hoofs are narrow and pointed, and a metatarsal gland, indicated by a tuft of white hairs, is present on the hock.

The height at the shoulder is 3 to 3½ feet, and the weight ranges from 150 to 300 pounds in males and 100 to 150 pounds in females.

The skull is large (more than 10 inches in length). A pit is present in the lacrimal bone, and the posterior nares are separated by the median vomer bone. The male in season is antlered, and each antler has one main beam, with supplemental branches or spikes, when present, directed backward. Dental formula: I 0/3, C 0/1, Pm 3/3, M 3/3.

Life History.—The white-tailed deer is most at home in woods and thickets, particularly where wooded areas alternate with open glades, fig. 1. By opening up the forests, early Illinois settlers increased the area favorable to the white-tail, and for a few years in the first half of the last century the deer population of the state increased. Later, more extensive destruction of forests, intensive agriculture, and hunting combined to exterminate the white-tail from Illinois. Reintroduced into this state, the white-tail has found suitable habitat in reforested areas and in other areas that are not cultivated.

Each white-tail buck annually grows a set of antlers, which start to develop late in April or early in May and are covered with velvety skin until it is rubbed off in August or September. It is beneath this "velvet" that bone-forming materials produce the hard antlers. During the mating or rutting season, the antlers are at their strongest, the neck of the male swells, and

Fig. 110.—Fawn of white-tailed deer.

the mature and vigorous bucks may take unto themselves a harem of several does. Breeding takes place in November or December. Later, in December or in January, there is a resorption of the bony material of each antler near the skull, and the antlers are shed.

Fawns are born in May or early June after a gestation period of 7 months. A doe may produce one, two, or three young at a birth. In Minnesota the average number of young per pregnant female is about 1.3. The young are protectively colored with a reddish coat, dotted with white, fig. 110, and are nearly free of scent. Throughout the summer, males and females usually remain separate.

Fig. 111.—Male and female adults of white-tailed deer.

The white-tail feeds on a variety of plants, including trees, nibbling or browsing a little on one and then on another item. If the deer are too numerous in an area, they will remove all of the available food they can reach, and some of them may starve to death. In some places, deer do damage to gardens and to orchards.

Signs.—Prints, fig. 15, of both front and hind feet of adult deer are relatively sharp pointed and usually register well when the animal is walking or trotting. Tracks of running deer show prints of the clots or short hind toes on each foot, especially in mud. Tracks of fawns or immature animals usually can be distinguished from tracks of pigs, sheep, and goats by the more acute pointedness of the footprints.

Deer droppings are long, oval objects, commonly varying from one-half to more than an inch long; generally one end is tipped with a cone-shaped projection. There are usually 50 to 100 pellets in each set.

Distribution.—The white-tailed deer was once common in Illinois, but the native stock was exterminated during the last century. In the 1930's, a concerted program of re-establishing the species was undertaken, and in the winter of 1950–51 game biologists estimated that there were more than 3,075 deer in 68 of the 102 counties of the state (Pietsch 1954:12). The introduced deer apparently belong to the subspecies *Odocoileus virginianus borealis* Miller. The natural range of the white-tailed deer includes most of North America south of an imaginary line extending from Nova Scotia to southeastern British Columbia.

BISON BISON (Linnaeus)

Bison   **Buffalo***

Description.—The bison is the only member of the bovine family which was native to Illinois within historic times. This animal is about the size of domestic cattle, but it appears larger because of its woolly, shaggy fur, its shaggy "beard," and the hump between its shoulders. The short horns and split hoofs are black. The horns, fig. 108*d,* are ever-growing and not shed. A bull weighs nearly a ton, a cow about half a ton.

---

*Bison is preferable to *buffalo* to distinguish this mammal from the water buffalo and other buffaloes of Africa and Asia.

Former Distribution.—The bison, or buffalo as it is more frequently called, once occurred throughout the prairies of Illinois, but probably never so abundantly as on the plains west of the Mississippi River. Early explorers record the bison mostly from along river banks, probably because these men usually traveled by way of water courses. They referred to "extraordinary" numbers of this animal and wrote of the prairies "abounding" in buffaloes. Large herds to these explorers apparently were made up of 200 to 300 individuals, not of several thousands, as in the West.

Numerous buffalo trails crossed and recrossed the prairies of the Illinois country. Perhaps the best known were those from Vincennes, Indiana, which invaded the lush central prairies from the east and continued on almost directly westward toward the Mississippi River.

By 1814, the bison had entirely disappeared from the Illinois territory, leaving behind their countless well-trodden paths.

At present, bison are kept in a semidomesticated condition on a few farms in Illinois.

SOME MAMMALS OF PREHISTORIC TIMES

Many thousands of years ago, there lived in the region of what is now known as Illinois certain kinds of mammals that became extinct long before the white man arrived in the Americas. These were the strange mammals of the Ice Age, or Pleistocene. Some flourished when the climate was cold, others when

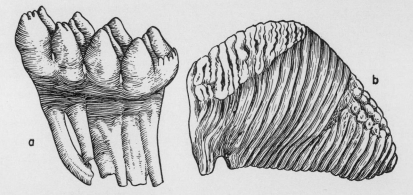

Fig. 112.—Tooth of a mastodon, *a*; tooth of a mammoth, *b*.

it was warm. Most notable of these animals were the elephant-like mammoths and mastodon, the musk-ox, the giant beaver, and the giant ground sloth. Their fossilized bones and teeth have

Fig. 113.—A restoration of the American mastodon.

Fig. 114.—A restoration of the mammoth.

been found in Illinois, and it is only through such evidence that their former presence in this area is known. Teeth of two of these prehistoric mammals, the mastodon and the mammoth, are pictured in fig. 112.

MEGALONYX JEFFERSONII Desmarest
Giant Ground Sloth

The most bizarre of all of the extinct mammals of the Illinois region was the giant ground sloth, which was nearly as large as a present-day elephant, but looked somewhat like a bear with a long neck and with a very broad, blunt tail. It apparently fed on leaves and twigs. Since sloths of today live in tropical America, it is usually assumed that the giant ground sloth occurred in the Illinois region during the warm interglacial periods. There are few fossil records of this prehistoric species for Illinois.

CASTOROIDES OHIOENSIS Foster
Giant Beaver

A gigantic beaver, as large as a bear, once frequented the waterways of the present-day Illinois area. One can surmise that the giant beaver constructed enormous dams and was capable of felling large trees. It became extinct at the close of the Ice Age. Several fossilized bones of this animal have been discovered in Illinois.

MAMMUT AMERICANUS (Kerr)
American Mastodon

The American mastodon, fig. 113, was about the size of an average elephant of present-day India, but the body was covered with long, coarse hair. The upper tusks were more or less parallel and either straight or curved upward. In the male, a single, short lower tusk was present, presumably completely concealed in the lower lip. The mastodon was probably a browser, feeding heavily upon leaves, for it lived in forested areas and had teeth for grinding food, fig. 112. Remains of this species have been found in a number of widely scattered localities in Illinois.

MAMMUTHUS JEFFERSONI (Osborn) and
MAMMUTHUS PRIMIGENIUS (Blumenbach)
Mammoth

These mammoths, fig. 114, were similar to but larger than present-day elephants. They stood approximately 11 feet high and had huge tusks, as long as 16 feet, bowed outward in the middle and curved inward at the tips. Their woolly fur, protected by long guard hairs, equipped them to withstand cold weather, and it is probable that they were in the Illinois region during periods of glaciation. Since their teeth, fig. 112, are so similar to those of present-day elephants, it is assumed that their food habits were much the same. Remains of two species of mammoths have been taken at scattered localities in Illinois.

EQUUS sp.
Horse

A few remains of a horse that was apparently somewhat like our present-day horse have been found in Illinois. Teeth of this prehistoric horse resemble those of our present-day horse, fig. 109c, but they can be distinguished by experts.

PLATYGONUS COMPRESSUS Le Conte
Peccary

The remains of a piglike animal, known as a peccary, have been taken near Galena and Alton, Illinois. The nearest relatives of this animal now live in southwestern United States and Mexico.

SYMBOS CAVIFRONS (Leidy)
Musk-Ox

During glacial periods the musk-ox occurred in what is now Illinois, and as far southward as Arkansas. Closely related musk-oxen are presently known from the barren Arctic of North America. The extinct musk-ox undoubtedly had long, shaggy fur, as do the living kinds. The horns were united over the head and extended downward before curving outward. There are very few records of this species in Illinois.

BISON sp.

Giant Bison   **Royal Bison**

A bison, fig. 115, much larger than the one that now exists, is known from skeletal remains found near Alton.

Fig. 115.—A restoration of the giant or royal bison.

RANGIFER sp.

Caribou

A kind of caribou is known from fossil remains found in Kentucky and near Alton, Illinois. Two kinds of caribou now occur in northern North America.

MAMMAL HABITATS AND THE FUTURE

Most of the mammalian species in the Illinois country when its settlement began are still with us. But, with few exceptions, they are here despite a lack of favorable attention paid to them. Some, like the thirteen-lined ground squirrel, have been inadvertently favored by agricultural practices, in this case excessive pasturing; some, like the cottontail rabbit, have been able to "take it"; others, like certain bats, have not been adversely affected by civilization; and a few, like the bobcat, have been able to survive because agriculture has by-passed habitats of varying sizes in which they can live.

The red squirrel, so far as we know, is the species most recently eliminated from the state. Nearer the center of its range, in cooler climates, this squirrel can tolerate a wide variety of habitats, some of them marginal. As far south as Illinois, only the better habitats are tolerable and, to be tolerable, they must be extensive.

Species other than the red squirrel have felt the pressure of Illinois agriculture. They have yielded ground as their habitats have been narrowed or eliminated. Most Illinois counties at one time had many species, widely distributed. Some counties now lack one or more species found in other parts of the state, and the species still present generally are restricted to a few small habitats. Gray squirrels, once widely distributed in all counties, now occcur only in a few choice habitats in most of those counties in which they are still present.

The loss of any one or several species may not, in itself, be of great consequence, but, when a species becomes rare or disappears, it serves as a signal that there has disappeared with it a vast and real, though often intangible, set of natural conditions which supported it. When bison disappeared from Illinois, they signaled the disappearance of extensive prairie, which was composed of native grasses, forbs, vines, and bushes.

We can now hardly afford in Illinois the extensive space required to maintain bison on natural habitats; elimination of bison from this state was a calculated procedure. Probably most people do not need the occasional solitude that a sojourn on the primitive prairie would afford; they experience sufficient contact with nature when they fish on a nearby lake, hunt on a farm, play a round of golf on a well-kept course, or attend a picnic in a public park. We are, however, in danger of letting too much of the natural habitat disappear, of crowding our satisfactions into too little out-of-door space.

In spite of decades of progressive soil conservation, too many Illinois pastures are still so heavily grazed that they not only do not support wildlife but they do not support the number of cattle they could if efficiently managed. Too many woodlands have cattle turned into them and thus lose many of their wild mammals and birds, and also their value as woods without attaining value as pastures, fig. 116. In 1949, two-thirds of the privately owned Illinois woodland was still being pastured: the equivalent of six counties going to waste! In such woodland, cattle use most of their energy in getting food; there is not enough food value in the meager grass to produce much milk and beef.

Too many ditchbanks, stream sides, and slopes are still farmed so severely that soil washes downstream with every rain. Too much water is allowed to run to sea—water that should sink deep into our soils to form a continuous store for human use. Too many brooks and rivers are fouled by oil and sewage, chemical waste and soil. Modern science can use or dispose of stream pollutants in harmless ways. Contour farming on slopes, restoration of plant cover to gullies and ravines, can hold

Fig. 116.—Left, managed, unpastured woodland; right, pastured woodland. Pastured woodland is uneconomical because it produces poor forage and little or no replacement timber.

soil in place and support game besides. What multiple rewards for small corrections!

Too much wildlife cover and too much wildlife have been destroyed by careless use of herbicides and insecticides; proper use of these materials requires experience, planning, and expert supervision.

Some corrections in the conservation situation have been made; they are bright pilots for the future. About a quarter century ago, a national forest was established in southern Illinois. In many parts of Illinois, state parks and county forests have been established. In heavily populated Cook County, a large forest preserve has for many years given reasonable opportunity for people to enjoy a soothing day in prairie and woodland areas. There are now in Illinois about 55 square miles of land in 64 state-owned parks and conservation areas, all of which serve as habitats for many species of wildlife. Railroad and highway rights-of-way harbor more species, usually the kinds that require grass or forb-covered habitats. These areas can be altered to harbor many more kinds of wildlife. In southern Illinois much of the land which was despoiled by ruinous farming is so poor that now, ironically, it is allowed, even helped, to produce timber without interference by grazing.

Mammalian populations renew themselves in such favorable centers and extend their ranges as more habitats become available. Within the past half century we have seen beavers and white-tailed deer restored in considerable numbers to Illinois, and furbearing mammals, with the possible exception of muskrats, are now more numerous in the state than they were 25 years ago. This favorable trend only indicates the possibility for the future. We can have more productive land *and* more wildlife. But there still is a long way to go.

GLOSSARY

albinistic. Whitish, having less than normal amount of pigment.

alveolar. Pertaining to the alveoli, as those of the jaws.

alveolus (pl. alveoli). A pit or socket, as for a tooth.

anterior. At or toward the head or front end; opposed to posterior.

antorbital canal. Canal in the skull having a small opening (infra-orbital foramen) to the outside in front of and usually slightly below the eye socket, fig. 117 (dark area above third premolar of upper jaw).

auditory bulla (pl., auditory bullae). Bony capsule (bulblike structure), fig. 117, at base of braincase surrounding the bones of the inner ear.

ball pad. A cushion-like thickening of the ball of the foot; corresponds to the ball of the human foot.

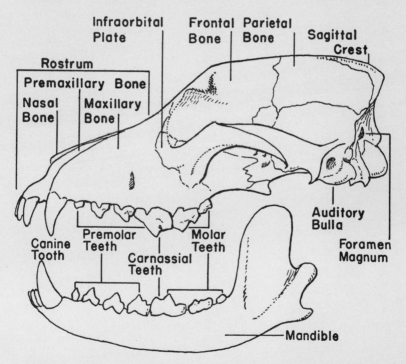

Fig. 117.—Side view of skull and lower jaw of dog. This and the following two figures are labeled to illustrate the diagnostic parts referred to in the keys and descriptions.

basioccipital bone. Bone at the base of the cranium (braincase) immediately in front of the foramen magnum (large opening at the base of the skull through which the spinal cord passes).

bony palate. See hard palate.

braincase. Part of the skull enclosing the brain, fig. 118.

bulla (pl., bullae). A hollow, thin-walled, bony prominence of rounded form, fig. 117.

calcar. In bats, a bony process (extension) from the heel; it partially supports the tail membrane, fig. 39.

canine or canine tooth. The tooth between the incisors (front cutting teeth) and the premolars; figs. 117, 119; eyetooth.

carnassial or carnassial tooth. Shearing tooth found in Carnivora; last upper premolar and first lower molar, figs. 117, 119.

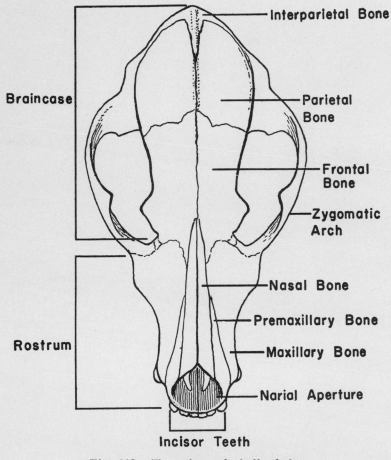

Fig. 118.—Top view of skull of dog.

carnivore. A mammal belonging to the order Carnivora.

carnivorous. Flesh-eating.

cheek pouch. Pouch either just inside or outside the mouth for the temporary storing of food.

cheek teeth. The premolar and molar teeth, fig. 119.

cranial. Of or pertaining to the skull, or to the cranium (braincase) only.

cranium. Braincase, fig. 118.

crown. Part of tooth not covered by the gum in living animal.

cusps. Prominences or points on the grinding surface of the crown of a tooth, fig. 112a.

dental formula. The arrangement of the teeth (for different species) written as a formula; example: I 3/3, C 1/1, Pm 4/4, M 2/3. The example should be interpreted as follows: To the

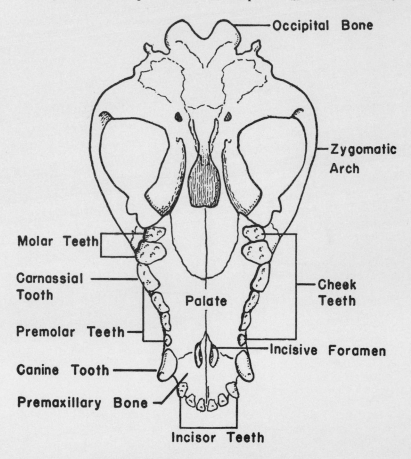

Fig. 119.—Under side of skull of dog.

left of each diagonal line is indicated the number of a certain kind of tooth in each half of the upper jaw, to the right the number of the same kind of tooth in each half of the lower jaw. The total number of teeth in each half of the upper jaw is 10, in each half of the lower jaw 11. Therefore, there are 20 upper teeth and 22 lower teeth and 42 teeth altogether. In each half of the upper jaw in this example there are three incisors (I), one canine (C), four premolars (Pm), and two molars (M).

dentary bone. One of a pair of membranous bones of the distal (farthest from hinge) part of the lower jaw.

diastema. A distinct space or gap between two teeth, as shown in fig. 40*f*.

digitigrade. Walking primarily on the toes, as in horses and dogs.

dorsal. Pertaining to the back or upper parts; opposed to ventral.

dorsoventrally. Between the dorsal (upper) and ventral (lower) sides of an animal. For example, a body or part flattened dorsoventrally would be unusually broad for its depth, i.e., vertically compressed as opposed to laterally compressed.

enamel loop. Loop formed by convolutions of enamel of tooth as seen on the grinding surface, fig. 74*a–e*.

enamel triangle. Triangle formed by convolutions of enamel of tooth as seen on the grinding surface, fig. 74*b, d, e*.

fenestrated. Having numerous openings, as in skull of a cottontail, fig. 40*d*.

flange. A rim or rimlike extension.

foramen. An opening, as found in the skull, through which nerves or blood vessels pass, fig. 119.

foramen magnum. Large opening in skull at posterior (rear) end, fig. 117.

forb. Any herbaceous plant other than a grass.

frontal or **frontal bone.** One of a pair of bones forming the roof of the braincase, figs. 117, 118.

fulvous. A dull yellowish gray or brownish color.

gestation period. The period of carrying of young normally in the uterus, from conception to delivery; period of pregnancy.

guard hairs. Long hairs of the pelt that protect the shorter and finer textured fur.

habitat. The natural abode in which a species of plant or animal grows or lives.

hard palate. Portion of palate, fig. 119, supported by bone; that part of palate covering the palatal bone.

herbivore. A mammal that feeds on plant material.

herbivorous. Plant eating.

hibernate. To pass the winter in close quarters in a torpid or inactive state.

incisive foramen. One of a pair of openings in the anterior palate, sometimes called anterior palatine foramen, figs. 73*d(f)*, 119.

incisor. One of the front teeth between the canines; found in the premaxillary or dentary bones, figs. 118, 119.

inferior. Lower, as opposed to superior or upper.

infraorbital foramen. A small opening or pore in the skull in front of, or anterior to, and usually slightly below the eye socket, fig. 117 (dark area above the third premolar of upper jaw).

infraorbital plate. Also called zygomatic plate; expanded anterior or front portion of the zygomatic arch, fig. 117.

insectivore. A mammal belonging to the order Insectivora (shrews and moles).

insectivorous. Insect eating.

interorbital region. Area between orbits or eye sockets in the skull.

interorbital space. Area across skull at the widest part of frontal bone.

interparietal or **interparietal bone.** Part of the top of the braincase; bone at the rear of the skull with the front extension wedged between the parietal bones, fig. 118.

lacrimal or **lacrimal bone.** Small bone at anterior (front) angle of eye socket.

lateral or **laterally.** Pertaining to the sides of an animal.

longitudinal or **longitudinally.** Lengthwise, along the long axis of an animal.

mandible. One of the halves of the lower jaw, fig. 117.

marsupial. Mammal belonging to the order Marsupialia; pertaining to an animal having a pouch for carrying the young.

maxilla (pl. **maxillae**) or **maxillary bone.** One of a pair of bones that bear the molar and premolar teeth, figs. 117, 118.

maxillary. Pertaining to the maxilla; maxillary bone.

melanistic. Black or blackish, having more than the normal amount of dark pigment.

metatarsal gland. A gland situated on the inside of the hind leg between the toes and the heel.

milk tooth. A temporary (deciduous) tooth; "baby" tooth.

molar or **molar tooth.** One of the posteriormost or back teeth of the upper or the lower jaw, figs. 74a–e, 117, 119.

molar tooth row. The group or row of molar teeth.

nares. Bones of the nose.

narial aperture. Internal nares; nasal opening through the roof of the mouth, fig. 118.

nasal or **nasal bone.** One of a pair of bones in the region of the nose, figs. 117, 118.

nosepad. The bare part of the nose of a mammal.

occipital or **occipital bone.** Pertaining to, or one of, the bones which surround the foramen magnum (large opening) at the posterior (rear) end of the skull, fig. 119.

ochraceous (**ocherous**). Pertaining to a color similar to ocher; referring to a yellowish or yellowish tan color.

omnivorous. Eating both animal and vegetable food.

orbit. Socket in the skull in which one of the eyes is located.

orbital region. Pertaining to the area near one of the orbits or eye sockets or between the eye sockets.

over-all length. Measurement from tip of the nose to tip of the tail, excluding terminal hairs; total length.

palatal width. Width of palate between teeth of the two sides of upper jaw, fig. 61*c, d.*

palate. Bony roof of the mouth, figs. 61*c, d,* 119.

parietal or parietal bone. One of a pair of bones of the braincase, figs. 117, 118.

pelage. The hair (fur or wool) covering the body of a mammal.

placenta. Organ of the developing embryo to provide for its nourishment and the elimination of waste products by interchange through the membranes of the uterus of the mother.

placental. Pertaining to an animal having a placenta.

plantigrade. Walking on the sole of the foot with the heel touching the ground, as in bears and man.

posterior. At or toward the hind end; opposed to anterior.

postorbital. Area situated behind the eye and usually referring to the postorbital process.

postorbital process. A projection of the frontal bone directly behind the orbit, fig. 73*b*(*p*).

predator. An animal that lives by preying upon other animals.

prehensile. Adapted for seizing or grasping.

premaxilla (pl. premaxillae). One of a pair of bones in the skull; these bones, between the nasal and maxillary bones, bear the upper incisor teeth, fig. 117; premaxillary bone, figs. 118, 119.

premaxillary. Pertaining to the premaxilla.

premolar. One of a series of teeth on each side of upper jaw and of lower jaw in front of the true molars, figs. 117, 119. When a canine tooth is present, the premolars are between it and the first molar, and in the upper jaw are confined to the maxillary bone. Premolars replace the milk (deciduous) molars.

process. A prominence or projecting part; an extension.

protuberant. Bulged beyond the surrounding or adjacent surface.

retractile. Capable of being drawn back or in, as with the claws of cats.

rostral. Pertaining to the rostrum or snout.

rostrum. The projecting front part of the skull, in the snout or nose region, figs. 40*c–d,* 44, 73*c–d,* 74*f–i,* 117, 118.

rudimentary. Not fully developed, or represented only by a vestige.

sagittal crest. Raised ridge of bone on top of and extending from front to back of braincase, fig. 117.

scat. A dropping of excrement.

septum. A partition.

| Gray | Keen's | Eastern | Red |
| Bat | Bat | Pipistrel | Bat |

skull length. Length of the skull as measured from the posterior-most portion to the anteriormost portion, fig. 118.

skull width. Width of the skull, usually the width across the zygomatic arches, fig. 118.

sonar system. A means of locating objects by producing sound waves and then receiving the reflected vibrations or sound waves.

supersonic. Pertaining to vibrations and waves whose frequencies are greater than those that affect the human ear.

supraorbital process. A process on the top rim of each orbit (eye socket), as in fig. 103.

supraorbital ridge. A lateral, longitudinal ridge of the skull situated above each eye socket.

suture. A line of union between bones of the skull.

tarsus. Part of the foot immediately below the long leg bones; the ankle.

temporal ridge. One of ridges on top of braincase, as in fig. 60*a, b*.

tentacle. An elongate, flexible, single or branched process (structure) usually tactile in function.

thoracic. Pertaining to the chest region.

tine. Any of the branches which come off the main beam of an antler.

tooth triangle. Same as enamel triangle.

tragus. Projection of skin near the bottom of the external ear opening in bats, fig. 51.

tubercle. A small knotlike prominence either soft or hard.

underfur. Thick, soft fur lying beneath the longer and coarser hair of a mammal.

ungulate. Hoofed; a hoofed animal.

unicuspid. A tooth with a single cusp, as in certain teeth of shrews, fig. 44*a, b, c.*

venter. Abdomen or belly.

ventral. Pertaining to the lower side, as opposed to dorsal.

vestigial. Pertaining to a structure or organ that in the adult is degenerate or imperfectly developed.

vomer or **vomer bone.** Bone forming part of the nasal septum (partition) and dividing the nasal cavities.

vomerine partition. Same as vomer.

zygomatic arch. An arch of bone forming the outside of each orbit (eye socket), figs. 118, 119.

zygomatic width. Width of skull across the zygomatic arches at the widest point.

Evening
Bat

Silver-Haired
Bat

Hoary
Bat

BIBLIOGRAPHY

Some Literature on Illinois Mammals

Anderson, Elsie P.
1951. The mammals of Fulton County, Illinois. Chicago Acad. Sci. Bul. 9(9):153–88.

Anonymous
1837. Illinois in 1837; a sketch descriptive of the situation, boundaries, face of the country, prominent districts, prairies, rivers, minerals, animlas [sic], agricultural productions, public lands, places of internal improvement, manufactures, &c. of the State of Illinois. S. Augustus Mitchell, Philadelphia. 143 pp.

Brown, Louis G., and Lee E. Yeager
1943. Survey of the Illinois fur resource. Ill. Nat. Hist. Surv. Bul. 22(6):435–504.

Cory, Charles B.
1912. The mammals of Illinois and Wisconsin. Field Mus. Nat. Hist. Pub. 153, Zool. Ser., 11:1–505.

Gregory, Tappan
1936. Mammals of the Chicago region. Chicago Acad. Sci. Prog. Act. 7(2–3):13–75.

Kennicott, Robert
1855. Catalogue of animals observed in Cook County, Illinois. Ill. State Ag. Soc. Trans. 1:577–95.
1857– The quadrupeds of Illinois injurious and beneficial to the
1859. farmer. U. S. Commr. Patents Rep. (Ag.) for 1856:52–110; 1857:72–107; 1858:241–56.

Mohr, Carl O.
1943. Illinois furbearer distribution and income. Ill. Nat. Hist. Surv. Bul. 22(7):505–37.

Necker, Walter L., and Donald M. Hatfield
1941. Mammals of Illinois. Chicago Acad. Sci. Bul. 6(3):17–60.

Pietsch, Lysle R.
1954. White-tailed deer populations in Illinois. Ill. Nat. Hist. Surv. Biol. Notes 34. 24 pp.

Sanborn, Colin C.
1925. Mammals of the Chicago area. Field Mus. Nat. Hist. Leaflet 8:129–51.

Sanborn, Colin Campbell, and Douglas Tibbitts
1949. Hoy's pygmy shrew in Illinois. Chicago Acad. Sci., Nat. Hist. Misc. 36. 2 pp.

Scott, Thomas G.
1955. An evaluation of the red fox. Ill. Nat. Hist. Surv. Biol. Notes 35. 16 pp.

Thomas, Cyrus
1861. Mammals of Illinois. Catalogue. Ill. State Ag. Soc. Trans. (for 1859–60) 4:651–61.

West, James A.
    1910. A study of the food of moles in Illinois. Ill. Lab. Nat.
           Hist. Bul. 9(2):14–22.
Wood, Frank Elmer
    1910. A study of the mammals of Champaign County, Illinois.
           Ill. Lab. Nat. Hist. Bul. 8:501–613.
Yeager, Lee E.
    1945. Capacity of Illinois land types to produce furs. N. Am.
           Wildlife Conf. Trans. 10:79–86.
Yeatter, Ralph E., and David H. Thompson
    1952. Tularemia, weather, and rabbit populations. Ill. Nat.
           Hist. Surv. Bul. 25(6):351–82.

Some Handbooks on Mammals of the United States

Anthony, H. E.
    1928. Field book of North American mammals. G. P. Putnam's
           Sons, New York and London. xxvi + 674 pp.
Burt, William Henry
    1952. A field guide to the mammals. Houghton Mifflin Com-
           pany, Boston. xxi + 200 pp.
Hamilton, William J., Jr.
    1943. The mammals of eastern United States. Comstock Pub-
           lishing Co., Inc., Ithaca, New York. 432 pp.
Murie, Olaus J.
    1954. A field guide to animal tracks. Houghton Mifflin Com-
           pany, Boston. xxii + 374 pp.
Palmer, Ralph S.
    1954. The mammal guide. Doubleday & Company, Inc., Garden
           City, New York. 384 pp.

Some Literature on Mammals of States Near Illinois

Barger, N. R.
    1951. Wisconsin mammals. Wis. Cons. Dept. Pub. 351–51. 54 pp.
Bennitt, Rudolf, and Werner O. Nagel
    1937. A survey of the resident game and furbearers of Mis-
           souri. Mo. Univ. Studies 12(2):1–215.
Burt, William H.
    1946. The mammals of Michigan. University of Michigan
           Press, Ann Arbor. xv + 288 pp.
Funkhouser, W. D.
    1925. Wild life in Kentucky. Ky. Geol. Surv., Ser. 6, 16:1–385.
Gunderson, Harvey L., and James R. Beer
    1953. The mammals of Minnesota. University of Minnesota
           Press, Minneapolis. xii + 190 pp.
Hahn, Walter Louis
    1907. Notes on mammals of the Kankakee Valley. U. S. Natl.
           Mus. Proc. 32:455–64.
    1909. The mammals of Indiana. Ind. Dept. Geol. and Nat.
           Res. Ann. Rep. 33:417–654.

Hicks, Ellis A., and George O. Hendrickson
    1940. Fur-bearers and game mammals of Iowa. Iowa State Col.
        Ag. Exp. Sta. Bul. P3(n.s.) :113–47.
Lyon, Marcus Ward, Jr.
    1936. Mammals of Indiana. Am. Midland Nat. 17(1):1–384.
Scott, Thos. G.
    1937. Mammals of Iowa. Iowa State Col. Jour. Sci. 12(1):43–98.
Scott, Thomas G., and Willard D. Klimstra
    1955. Red foxes and a declining prey population. So. Ill. Univ.
        Monog. Ser. 1. 123 pp.
Seagears, Clayton B.
    1945. The fox in New York (second printing). New York State
        Conservation Department, Albany. 85 pp.
Swanson, Gustav, Thaddeus Surber, and Thomas S. Roberts
    1945. The mammals of Minnesota. Minn. Dept. Cons. Tech.
        Bul. 2. 108 pp.

INDEX

Page entries in **boldface** type refer to principal treatments of items listed; those in *italic* type refer to illustrations. Names of orders and families are in CAPITAL letters; those of genera and species are in *italic* type. All names of animals and certain other words are listed in the index in the singular, as mouse for mice, tooth for teeth, regardless of the way in which they appear in the text. For complete reference to an animal, pages under both scientific and common names should be consulted.

A CATALOGUE OF SELECTED DOVER BOOKS
IN ALL FIELDS OF INTEREST

A CATALOGUE OF SELECTED DOVER BOOKS
IN ALL FIELDS OF INTEREST

AMERICA'S OLD MASTERS, James T. Flexner. Four men emerged unexpectedly from provincial 18th century America to leadership in European art: Benjamin West, J. S. Copley, C. R. Peale, Gilbert Stuart. Brilliant coverage of lives and contributions. Revised, 1967 edition. 69 plates. 365pp. of text.
21806-6 Paperbound $3.00

FIRST FLOWERS OF OUR WILDERNESS: AMERICAN PAINTING, THE COLONIAL PERIOD, James T. Flexner. Painters, and regional painting traditions from earliest Colonial times up to the emergence of Copley, West and Peale Sr., Foster, Gustavus Hesselius, Feke, John Smibert and many anonymous painters in the primitive manner. Engaging presentation, with 162 illustrations. xxii + 368pp.
22180-6 Paperbound $3.50

THE LIGHT OF DISTANT SKIES: AMERICAN PAINTING, 1760-1835, James T. Flexner. The great generation of early American painters goes to Europe to learn and to teach: West, Copley, Gilbert Stuart and others. Allston, Trumbull, Morse; also contemporary American painters—primitives, derivatives, academics—who remained in America. 102 illustrations. xiii + 306pp. 22179-2 Paperbound $3.00

A HISTORY OF THE RISE AND PROGRESS OF THE ARTS OF DESIGN IN THE UNITED STATES, William Dunlap. Much the richest mine of information on early American painters, sculptors, architects, engravers, miniaturists, etc. The only source of information for scores of artists, the major primary source for many others. Unabridged reprint of rare original 1834 edition, with new introduction by James T. Flexner, and 394 new illustrations. Edited by Rita Weiss. 6⅝ x 9⅜.
21695-0, 21696-9, 21697-7 Three volumes, Paperbound $13.50

EPOCHS OF CHINESE AND JAPANESE ART, Ernest F. Fenollosa. From primitive Chinese art to the 20th century, thorough history, explanation of every important art period and form, including Japanese woodcuts; main stress on China and Japan, but Tibet, Korea also included. Still unexcelled for its detailed, rich coverage of cultural background, aesthetic elements, diffusion studies, particularly of the historical period. 2nd, 1913 edition. 242 illustrations. lii + 439pp. of text.
20364-6, 20365-4 Two volumes, Paperbound $6.00

THE GENTLE ART OF MAKING ENEMIES, James A. M. Whistler. Greatest wit of his day deflates Oscar Wilde, Ruskin, Swinburne; strikes back at inane critics, exhibitions, art journalism; aesthetics of impressionist revolution in most striking form. Highly readable classic by great painter. Reproduction of edition designed by Whistler. Introduction by Alfred Werner. xxxvi + 334pp.
21875-9 Paperbound $2.50

INCIDENTS OF TRAVEL IN YUCATAN, John L. Stephens. Classic (1843) exploration of jungles of Yucatan, looking for evidences of Maya civilization. Stephens found many ruins; comments on travel adventures, Mexican and Indian culture. 127 striking illustrations by F. Catherwood. Total of 669 pp.
20926-1, 20927-X Two volumes, Paperbound $5.00

INCIDENTS OF TRAVEL IN CENTRAL AMERICA, CHIAPAS, AND YUCATAN, John L. Stephens. An exciting travel journal and an important classic of archeology. Narrative relates his almost single-handed discovery of the Mayan culture, and exploration of the ruined cities of Copan, Palenque, Utatlan and others; the monuments they dug from the earth, the temples buried in the jungle, the customs of poverty-stricken Indians living a stone's throw from the ruined palaces. 115 drawings by F. Catherwood. Portrait of Stephens. xii + 812pp.
22404-X, 22405-8 Two volumes, Paperbound $6.00

A NEW VOYAGE ROUND THE WORLD, William Dampier. Late 17-century naturalist joined the pirates of the Spanish Main to gather information; remarkably vivid account of buccaneers, pirates; detailed, accurate account of botany, zoology, ethnography of lands visited. Probably the most important early English voyage, enormous implications for British exploration, trade, colonial policy. Also most interesting reading. Argonaut edition, introduction by Sir Albert Gray. New introduction by Percy Adams. 6 plates, 7 illustrations. xlvii + 376pp. 6½ x 9¼.
21900-3 Paperbound $3.00

INTERNATIONAL AIRLINE PHRASE BOOK IN SIX LANGUAGES, Joseph W. Bátor. Important phrases and sentences in English paralleled with French, German, Portuguese, Italian, Spanish equivalents, covering all possible airport-travel situations; created for airline personnel as well as tourist by Language Chief, Pan American Airlines. xiv + 204pp.
22017-6 Paperbound $2.00

STAGE COACH AND TAVERN DAYS, Alice Morse Earle. Detailed, lively account of the early days of taverns; their uses and importance in the social, political and military life; furnishings and decorations; locations; food and drink; tavern signs, etc. Second half covers every aspect of early travel; the roads, coaches, drivers, etc. Nostalgic, charming, packed with fascinating material. 157 illustrations, mostly photographs. xiv + 449pp.
22518-6 Paperbound $4.00

NORSE DISCOVERIES AND EXPLORATIONS IN NORTH AMERICA, Hjalmar R. Holand. The perplexing Kensington Stone, found in Minnesota at the end of the 19th century. Is it a record of a Scandinavian expedition to North America in the 14th century? Or is it one of the most successful hoaxes in history. A scientific detective investigation. Formerly *Westward from Vinland.* 31 photographs, 17 figures. x + 354pp.
22014-1 Paperbound $2.75

A BOOK OF OLD MAPS, compiled and edited by Emerson D. Fite and Archibald Freeman. 74 old maps offer an unusual survey of the discovery, settlement and growth of America down to the close of the Revolutionary war: maps showing Norse settlements in Greenland, the explorations of Columbus, Verrazano, Cabot, Champlain, Joliet, Drake, Hudson, etc., campaigns of Revolutionary war battles, and much more. Each map is accompanied by a brief historical essay. xvi + 299pp. 11 x 13¾.
22084-2 Paperbound $6.00

ADVENTURES OF AN AFRICAN SLAVER, Theodore Canot. Edited by Brantz Mayer. A detailed portrayal of slavery and the slave trade, 1820-1840. Canot, an established trader along the African coast, describes the slave economy of the African kingdoms, the treatment of captured negroes, the extensive journeys in the interior to gather slaves, slave revolts and their suppression, harems, bribes, and much more. Full and unabridged republication of 1854 edition. Introduction by Malcom Cowley. 16 illustrations. xvii + 448pp. 22456-2 Paperbound $3.50

MY BONDAGE AND MY FREEDOM, Frederick Douglass. Born and brought up in slavery, Douglass witnessed its horrors and experienced its cruelties, but went on to become one of the most outspoken forces in the American anti-slavery movement. Considered the best of his autobiographies, this book graphically describes the inhuman treatment of slaves, its effects on slave owners and slave families, and how Douglass's determination led him to a new life. Unaltered reprint of 1st (1855) edition. xxxii + 464pp. 22457-0 Paperbound $2.50

THE INDIANS' BOOK, recorded and edited by Natalie Curtis. Lore, music, narratives, dozens of drawings by Indians themselves from an authoritative and important survey of native culture among Plains, Southwestern, Lake and Pueblo Indians. Standard work in popular ethnomusicology. 149 songs in full notation. 23 drawings, 23 photos. xxvi + 584pp. 6⅝ x 9⅜. 21939-9 Paperbound $4.50

DICTIONARY OF AMERICAN PORTRAITS, edited by Hayward and Blanche Cirker. 4024 portraits of 4000 most important Americans, colonial days to 1905 (with a few important categories, like Presidents, to present). Pioneers, explorers, colonial figures, U. S. officials, politicians, writers, military and naval men, scientists, inventors, manufacturers, jurists, actors, historians, educators, notorious figures. Indian chiefs, etc. All authentic contemporary likenesses. The only work of its kind in existence; supplements all biographical sources for libraries. Indispensable to anyone working with American history. 8,000-item classified index, finding lists, other aids. xiv + 756pp. 9¼ x 12¾. 21823-6 Clothbound $30.00

TRITTON'S GUIDE TO BETTER WINE AND BEER MAKING FOR BEGINNERS, S. M. Tritton. All you need to know to make family-sized quantities of over 100 types of grape, fruit, herb and vegetable wines; as well as beers, mead, cider, etc. Complete recipes, advice as to equipment, procedures such as fermenting, bottling, and storing wines. Recipes given in British, U. S., and metric measures. Accompanying booklet lists sources in U. S. A. where ingredients may be bought, and additional information. 11 illustrations. 157pp. 5⅝ x 8⅛.
(USO) 22090-7 Clothbound $3.50

GARDENING WITH HERBS FOR FLAVOR AND FRAGRANCE, Helen M. Fox. How to grow herbs in your own garden, how to use them in your cooking (over 55 recipes included), legends and myths associated with each species, uses in medicine, perfumes, etc.—these are elements of one of the few books written especially for American herb fanciers. Guides you step-by-step from soil preparation to harvesting and storage for each type of herb. 12 drawings by Louise Mansfield. xiv + 334pp.
22540-2 Paperbound $2.50

JIM WHITEWOLF: THE LIFE OF A KIOWA APACHE INDIAN, Charles S. Brant, editor. Spans transition between native life and acculturation period, 1880 on. Kiowa culture, personal life pattern, religion and the supernatural, the Ghost Dance, breakdown in the White Man's world, similar material. 1 map. xii + 144pp.
22015-X Paperbound $1.75

THE NATIVE TRIBES OF CENTRAL AUSTRALIA, Baldwin Spencer and F. J. Gillen. Basic book in anthropology, devoted to full coverage of the Arunta and Warramunga tribes; the source for knowledge about kinship systems, material and social culture, religion, etc. Still unsurpassed. 121 photographs, 89 drawings. xviii + 669pp.
21775-2 Paperbound $5.00

MALAY MAGIC, Walter W. Skeat. Classic (1900); still the definitive work on the folklore and popular religion of the Malay peninsula. Describes marriage rites, birth spirits and ceremonies, medicine, dances, games, war and weapons, etc. Extensive quotes from original sources, many magic charms translated into English. 35 illustrations. Preface by Charles Otto Blagden. xxiv + 685pp.
21760-4 Paperbound $4.00

HEAVENS ON EARTH: UTOPIAN COMMUNITIES IN AMERICA, 1680-1880, Mark Holloway. The finest nontechnical account of American utopias, from the early Woman in the Wilderness, Ephrata, Rappites to the enormous mid 19th-century efflorescence; Shakers, New Harmony, Equity Stores, Fourier's Phalanxes, Oneida, Amana, Fruitlands, etc. "Entertaining and very instructive." *Times Literary Supplement.* 15 illustrations. 246pp.
21593-8 Paperbound $2.00

LONDON LABOUR AND THE LONDON POOR, Henry Mayhew. Earliest (c. 1850) sociological study in English, describing myriad subcultures of London poor. Particularly remarkable for the thousands of pages of direct testimony taken from the lips of London prostitutes, thieves, beggars, street sellers, chimney-sweepers, street-musicians, "mudlarks," "pure-finders," rag-gatherers, "running-patterers," dock laborers, cab-men, and hundreds of others, quoted directly in this massive work. An extraordinarily vital picture of London emerges. 110 illustrations. Total of lxxvi + 1951pp. 6⅝ x 10.
21934-8, 21935-6, 21936-4, 21937-2 Four volumes, Paperbound $14.00

HISTORY OF THE LATER ROMAN EMPIRE, J. B. Bury. Eloquent, detailed reconstruction of Western and Byzantine Roman Empire by a major historian, from the death of Theodosius I (395 A.D.) to the death of Justinian (565). Extensive quotations from contemporary sources; full coverage of important Roman and foreign figures of the time. xxxiv + 965pp. 21829-5 Record, book, album. Monaural. $3.50

AN INTELLECTUAL AND CULTURAL HISTORY OF THE WESTERN WORLD, Harry Elmer Barnes. Monumental study, tracing the development of the accomplishments that make up human culture. Every aspect of man's achievement surveyed from its origins in the Paleolithic to the present day (1964); social structures, ideas, economic systems, art, literature, technology, mathematics, the sciences, medicine, religion, jurisprudence, etc. Evaluations of the contributions of scores of great men. 1964 edition, revised and edited by scholars in the many fields represented. Total of xxix + 1381pp. 21275-0, 21276-9, 21277-7 Three volumes, Paperbound $7.75

AMERICAN FOOD AND GAME FISHES, David S. Jordan and Barton W. Evermann. Definitive source of information, detailed and accurate enough to enable the sportsman and nature lover to identify conclusively some 1,000 species and sub-species of North American fish, sought for food or sport. Coverage of range, physiology, habits, life history, food value. Best methods of capture, interest to the angler, advice on bait, fly-fishing, etc. 338 drawings and photographs. 1 + 574pp. $6\frac{5}{8}$ x $9\frac{3}{8}$.

22383-1 Paperbound $4.50

THE FROG BOOK, Mary C. Dickerson. Complete with extensive finding keys, over 300 photographs, and an introduction to the general biology of frogs and toads, this is the classic non-technical study of Northeastern and Central species. 58 species; 290 photographs and 16 color plates. xvii + 253pp.

21973-9 Paperbound $4.00

THE MOTH BOOK: A GUIDE TO THE MOTHS OF NORTH AMERICA, William J. Holland. Classical study, eagerly sought after and used for the past 60 years. Clear identification manual to more than 2,000 different moths, largest manual in existence. General information about moths, capturing, mounting, classifying, etc., followed by species by species descriptions. 263 illustrations plus 48 color plates show almost every species, full size. 1968 edition, preface, nomenclature changes by A. E. Brower. xxiv + 479pp. of text. $6\frac{1}{2}$ x $9\frac{1}{4}$.

21948-8 Paperbound $5.00

THE SEA-BEACH AT EBB-TIDE, Augusta Foote Arnold. Interested amateur can identify hundreds of marine plants and animals on coasts of North America; marine algae; seaweeds; squids; hermit crabs; horse shoe crabs; shrimps; corals; sea anemones; etc. Species descriptions cover: structure; food; reproductive cycle; size; shape; color; habitat; etc. Over 600 drawings. 85 plates. xii + 490pp.

21949-6 Paperbound $3.50

COMMON BIRD SONGS, Donald J. Borror. $33\frac{1}{3}$ 12-inch record presents songs of 60 important birds of the eastern United States. A thorough, serious record which provides several examples for each bird, showing different types of song, individual variations, etc. Inestimable identification aid for birdwatcher. 32-page booklet gives text about birds and songs, with illustration for each bird.

21829-5 Record, book, album. Monaural. $2.75

FADS AND FALLACIES IN THE NAME OF SCIENCE, Martin Gardner. Fair, witty appraisal of cranks and quacks of science: Atlantis, Lemuria, hollow earth, flat earth, Velikovsky, orgone energy, Dianetics, flying saucers, Bridey Murphy, food fads, medical fads, perpetual motion, etc. Formerly "In the Name of Science." x + 363pp.

20394-8 Paperbound $2.00

HOAXES, Curtis D. MacDougall. Exhaustive, unbelievably rich account of great hoaxes: Locke's moon hoax, Shakespearean forgeries, sea serpents, Loch Ness monster, Cardiff giant, John Wilkes Booth's mummy, Disumbrationist school of art, dozens more; also journalism, psychology of hoaxing. 54 illustrations. xi + 338pp.

20465-0 Paperbound $2.75

MATHEMATICAL PUZZLES FOR BEGINNERS AND ENTHUSIASTS, Geoffrey Mott-Smith. 189 puzzles from easy to difficult—involving arithmetic, logic, algebra, properties of digits, probability, etc.—for enjoyment and mental stimulus. Explanation of mathematical principles behind the puzzles. 135 illustrations. viii + 248pp.
20198-8 Paperbound $1.75

PAPER FOLDING FOR BEGINNERS, William D. Murray and Francis J. Rigney. Easiest book on the market, clearest instructions on making interesting, beautiful origami. Sail boats, cups, roosters, frogs that move legs, bonbon boxes, standing birds, etc. 40 projects; more than 275 diagrams and photographs. 94pp.
20713-7 Paperbound $1.00

TRICKS AND GAMES ON THE POOL TABLE, Fred Herrmann. 79 tricks and games—some solitaires, some for two or more players, some competitive games—to entertain you between formal games. Mystifying shots and throws, unusual caroms, tricks involving such props as cork, coins, a hat, etc. Formerly *Fun on the Pool Table*. 77 figures. 95pp.
21814-7 Paperbound $1.00

HAND SHADOWS TO BE THROWN UPON THE WALL: A SERIES OF NOVEL AND AMUSING FIGURES FORMED BY THE HAND, Henry Bursill. Delightful picturebook from great-grandfather's day shows how to make 18 different hand shadows: a bird that flies, duck that quacks, dog that wags his tail, camel, goose, deer, boy, turtle, etc. Only book of its sort. vi + 33pp. 6½ x 9¼.
21779-5 Paperbound $1.00

WHITTLING AND WOODCARVING, E. J. Tangerman. 18th printing of best book on market. "If you can cut a potato you can carve" toys and puzzles, chains, chessmen, caricatures, masks, frames, woodcut blocks, surface patterns, much more. Information on tools, woods, techniques. Also goes into serious wood sculpture from Middle Ages to present, East and West. 464 photos, figures. x + 293pp.
20965-2 Paperbound $2.00

HISTORY OF PHILOSOPHY, Julián Marias. Possibly the clearest, most easily followed, best planned, most useful one-volume history of philosophy on the market; neither skimpy nor overfull. Full details on system of every major philosopher and dozens of less important thinkers from pre-Socratics up to Existentialism and later. Strong on many European figures usually omitted. Has gone through dozens of editions in Europe. 1966 edition, translated by Stanley Appelbaum and Clarence Strowbridge. xviii + 505pp.
21739-6 Paperbound $3.00

YOGA: A SCIENTIFIC EVALUATION, Kovoor T. Behanan. Scientific but non-technical study of physiological results of yoga exercises; done under auspices of Yale U. Relations to Indian thought, to psychoanalysis, etc. 16 photos. xxiii + 270pp.
20505-3 Paperbound $2.50

Prices subject to change without notice.
Available at your book dealer or write for free catalogue to Dept. GI, Dover Publications, Inc., 180 Varick St., N. Y., N. Y. 10014. Dover publishes more than 150 books each year on science, elementary and advanced mathematics, biology, music, art, literary history, social sciences and other areas.